Qualitative Research Using R: A Systematic Approach

Yanto Chandra • Liang Shang

Qualitative Research Using R: A Systematic Approach

Springer

Yanto Chandra
The Hong Kong Polytechnic University
Hong Kong, Kowloon, Hong Kong

Liang Shang
City University of Hong Kong
Hong Kong, Kowloon, Hong Kong

ISBN 978-981-13-3169-5 ISBN 978-981-13-3170-1 (eBook)
https://doi.org/10.1007/978-981-13-3170-1

Library of Congress Control Number: 2016963685

This Springer imprint is published by the registered company Springer Nature Singapore Pte Ltd.
The registered company address is: 152 Beach Road, #21-01/04 Gateway East, Singapore 189721, Singapore

Foreword

Years ago, I was facing a problem that has haunted qualitative researchers forever: How do you imbue inductive research with "qualitative rigor" and still hold on to its marvelous power for generating revelatory insights? I was frustrated in my attempts to publish qualitative research, because reviewers just didn't seem to "get" what I was trying to do. So, I puzzled over whether it might be possible to develop a systematic approach to doing qualitative research in a way that could convince reviewers, editors, and interested readers that my conclusions were not only insightful but plausible, defensible, and rooted in convincing data. Of course, I wanted to keep and even enhance qualitative research's great historical seductions: discovering new concepts, revealing deep structures and processes, fostering new ways of seeing, etc.

I also wanted to develop some way to counter the (then often accurate) perception that qualitative research was overly impressionistic and founded on slim evidence. In addition, I wanted to give voice to the people actually experiencing a phenomenon, rather than furthering the fiction that the researcher has some sort of theoretical omniscience. In my view, people doing things in organizations are usually knowledgeable (I know because I once was one of those people doing consequential things in a prominent organization). They generally have a good sense of what they are trying to do, how they are trying to do it, and why they are trying to do it – and they usually can explain it to intellectual eggheads in understandable terms. Perhaps most important of all, I wanted to be able to generate grounded theoretical explanations of their experience – explanations that were predicated on demonstrable data-to-theory connections, so that I could "prove" to skeptical gatekeepers that I knew what I was talking about. Big dreams with big hopes.

To accomplish these kinds of ambitious aims, we scholars are all looking for ways to convince readers of the value of our work, either with compelling data or persuasive rhetoric (or preferably both). My question was: Could I develop an approach that would not only guide my research but also convince readers that my findings were credible and rooted in verifiable evidence? On some practical level, I also wanted to avoid the dreaded RwS (Rejection with Scorn) that frequently came with editor exhortations to show the evidence for your assertions and to better demonstrate your data-to-theory connections.

My approach to fulfilling these grand ambitions and to addressing the perennial concerns of reviewers and editors has been to present both a first-order account (adequate at the level of meaning of the informants) and a second-order account (adequate at the level of meaning of the researchers and theorists). In my way of doing, this process generates a "data structure" that shows all the requisite data-to-theory connections, serves as a basis for writing the findings narrative, and provides the foundation for a grounded process model.

In the early days, I did all the manipulation and analysis of my textual data manually, with lots of index cards, color-coded "Post-It" notes, and myriad category organizing systems. I frequently commandeered conference rooms for days so I could lay everything out on a giant table and stand on that table to get a grand view of a complex and messy data universe. It was tedious in the extreme, but it also was a necessary avenue to revelation. Some years later, when qualitative data analysis programs like NVivo, ATLAS.ti, and MaxQDA came along, the tasks of pattern establishment and similarity/difference analysis could be streamlined by remanding the job to software. My thanks to all those mavens who made research life a lot less wearying.

Now, along come a couple of latter-day saviors, Yanto Chandra and Liang Shang, to make the qualitative analytical world so much easier to navigate for the well-intended, if overwhelmed, qualitative researcher. They have enhanced the approach I have been continuously developing for many years by bringing modern technology to bear on the data processing and evidence-generating aspects of the approach and, consequently, have freed up more brain space for the distinctively human talents for creativity and insight.

This book constitutes a valuable treatise for integrating my approach to qualitative research with computer-assisted qualitative data analysis software and the RQDA, which runs inside the (free!) R computing platform. It provides hold-you-by-the hand, step-by-step instructions and screen displays so that you can become your own qualitative research maven. Even if you have somehow forgotten how to count – not to worry – Chandra and Shang can even take care of that little problem for you. Their synthesis brings a systematic qualitative approach into the twenty-first century.

Robert and Judith Auritt Klein Professor of Management Dennis Gioia
Smeal College of Business at Penn State University,
University Park, PL, USA
25 May 2018

Endorsements

Qualitative Research Using R: A Systematic Approach guides aspiring researchers through the process of conducting a qualitative study with the assistance of the R programming language. It is the only textbook that offers "click-by-click" instruction in how to use RQDA software to carry out analysis. This book will undoubtedly serve as a useful resource for those interested in learning more about R as applied to qualitative or mixed methods data analysis. Helpful as well is the six-step procedure for carrying out a grounded theory-type study (the "Gioia approach") with the support of RQDA software, making it a comprehensive resource for those interested in innovative qualitative methods and uses of CAQDAS tools.

Trena M. Paulus, Professor of Education, University of Georgia, USA

Chandra and Shang have written a timely book that expands the methodological toolkit for qualitative research. This book takes the reader step by step through a computerized process for conducting qualitative research that is detailed, easy to follow and full of examples. Based on open-source software sensitive to the various paradigms underlying qualitative research, Chandra and Shang have made the process accessible to all qualitative researchers.

Trish Ruebottom, Professor, Goodman School of Business, Brock University, USA

Qualitative Research Using R: A Systematic Approach is a timely book which offers qualitative scholars a much-needed, easy-to-understand, and step-by-step introduction to RQDA, an open-source CAQDAS software for analyzing qualitative data. Rigorous and practical, the book by Yanto Chandra and Liang Shang is indispensable for any qualitative researcher who wants to acquaint themselves with the technicalities of RQDA and who wants to learn how RQDA can be used to perform state-of-the-art research that is equally transparent, trustworthy, and rigorous.

Pascal Dey, Professor, Grenoble Ecole de Management, France

For researchers who are looking for a DIY manual for structuring research, this book is a gateway into an exciting world of narrative research, all that an aspiring Ph.D. researcher needs in order to design and carry out a winning research design centering on narrative analysis.

Scott Victor Valentine, Professor and Associate Dean of Sustainability and Urban Planning, RMIT University, Australia

In their book, Chandra and Shang give a step-by-step tutorial for using R for qualitative data analysis. The key feature of this book is the way the authors demonstrate the use of R software using relevant examples. This book paves way for students and researchers to use this open-source, free software for qualitative research purposes. It literally brings cost-effective qualitative data analysis "to the masses."

Dr. Lakshmi B. Nair, Faculty of Social and Behavioural Sciences, Utrecht University, Netherlands

Preface

This book represents an evolution of over 15 years of labor, learning, and self-reinvention since the first author started tinkering with different technologies of rationality to enhance qualitative and quantitative research. This is the first author's response to having worked for multinational firms and, today, as an academic, author, and journal gatekeeper. Together with Shang Liang, my collaborator in the past few years, we have engaged in daily battles with gatekeepers in top-tier journals and understand the challenges of *persuading* others that the conclusion from our research is plausible, defensible, trustworthy, and transparent. How can we achieve this tall order using a research that primarily relies on textual data with no *p*-values or r-squares to "prove" our findings are sound? As scholars, how can we convince readers that we did observe concept A or event B or process C in the data and that our decisions in making sense of and giving sense to the data have evidentiary basis? How can gatekeepers be assured that we have met ape-like men in the forest of Sumatra by just saying "I saw them, I was there!"?

The short answer to these big questions is to *systematize* how qualitative research is designed, executed, written, and communicated to all stakeholders and to provide a trail of *evidence* to let others see how a conclusion is reached. Adopting a *systematic* approach to conducting qualitative research does not mean embracing naive empiricist-positivist research, as some might believe. In fact, we can be systematic and also creative, imaginative, and improvisational to attain the highest possible quality of qualitative research. Using a systematic approach is *creativity with fences*. It's about *responsible scholarship* in which we demonstrate – with evidence – how we start from an idea to reach a plausible and defensible conclusion while telling an interesting story.

That said, being unsystematic in doing qualitative research does not automatically qualify as high-quality interpretivist research. The work of a qualitative scholar resembles that of a lawyer. Like a lawyer, we may not know the actual or empirical truth of a case under investigation or even whether this truth exists or not, but we can, like lawyers, provide solid, verifiable, and convincing evidence, evidence that we have systematically collected, analyzed, and presented to support a plausible and defensible conclusion. The rest is up to the jury to decide.

Is it desirable to have a *consensus* for the qualitative research enterprise? Is it possible to achieve such consensus? This book is not an appropriate channel to debate consensus in qualitative research. We do agree, however, that there are *conventions* or *dominant* approaches to performing qualitative research. Like product battles in the commercial marketplace, there are battles of ideas in the academic marketplace, and the "winners" consequently promote the *dominant designs* or *paradigms* that influence the users. To date (as will be discussed in Chap. 1), we are witnessing the rise of three major paradigms in qualitative research – that of Gioia, Yin-Eisenhardt, and Langley. These scholars are key opinion leaders who are shaping the qualitative methods landscape. As discussed in this book, each approach is unique and has distinctive strengths and purposes. Although some qualitative scholars advocate methodological pluralism and reject the need for consensus as to what qualifies as good qualitative research, we fear that the pluralists themselves are creating another paradigm – a non-paradigmatic approach. If "a way of seeing is a way of not seeing," then plausibly, "a way of not seeing is a way of seeing" is also true. So, let's be honest and admit that we, as scholars, belong to and anchor our work to a paradigm, even if we deny it.

As authors, we consciously choose to take a *position* in the current paradigmatic and consensus debate (a debate about the importance of having a dominant approach or a set of procedures to execute it) in the qualitative research enterprise. In our research, we tend to work with in-depth interview data and other forms of textual data (e.g., profiles of social innovators/entrepreneurs from public sources, Facebook and Twitter posts, news articles), and we often have a large amount of data, ranging from dozens to tens of thousands. We do not always have access to data on the performance or outcomes of the phenomena we are studying (e.g., high- vs. low-performing cases). We cannot always access longitudinal data from the case sites we work with. Although we can study the social reality like the critical realists do (e.g., "map *events* and go backwards to infer multiple paths of causality") or like the empiricist-positivists do (e.g., "verify and confirm patterns in cases using qualitative and quantitative data"), we found that the questions we have asked and types of data we have worked with so far tend to lend ourselves to the interpretivist-constructivist approach. Thus, consciously or unconsciously, we have aligned ourselves with the *Gioia approach* – an approach that some scholars thought to be too rigid and positivistic.

Based on our experience, the Gioia approach is a powerful methodology to conduct high-quality interpretivist-constructivist qualitative research. But we also understand that the Gioia approach needs enhancement using technologies of rationality to demonstrate its systematic and evidentiary bases with a strong dose of creative flavor. As users, learners, and hobbyists of open-source (free) computing platforms such as R, we found that the Gioia approach can be fruitfully fused with R. The fusion leads to a new era for executing the Gioia approach and opens new pathways for researchers with no programming background as well as savvy R users to perform qualitative research in a new way assisted by a powerful software package in R called RQDA.

Hong Kong SAR, China

Yanto Chandra
Liang Shang

About This Book

In the language of qualitative scholarship, RQDA – R package for Qualitative Data Analysis – belongs to the computer-assisted qualitative data analysis software (CAQDAS). The growing interest in CAQDAS-based qualitative research is well reflected in the increasing number of books that address this topic. Examples include *technical-oriented* CAQDAS software textbooks such as those focusing on NVivo (Bazeley 2007; Richards 2014) and ATLAS.ti software (Friese 2011; Woolf and Silver 2017). The number of *general* qualitative research methods books (e.g., Maxwell 2013; Tracy 2012; Savin-Baden and Howell 2012) has also grown in recent years. What is missing, and what qualitative scholars most urgently need, is a book or manual that *integrates the methodological* and *technical content* for CAQDAS-based qualitative research. The world of R contains numerous books on R programming, and they all focus on the quantitative applications of R, but not one concentrates on qualitative research.

This book is the first that integrates CAQDAS, particularly RQDA and the Gioia approach, to offer step-by-step instructions on how to conduct qualitative research. Our book begins with a discussion on the rise of the Gioia approach as an important paradigm in social science qualitative research – in addition to the Yin-Eisenhardt and Langley approaches – and demonstrates how the Gioia approach can be operationalized and enhanced using an R package called RQDA. It provides a *methodological* and *technical* review of RQDA as a new CAQDAS tool. Using samples of textual data, we demonstrate the execution of the R-based Gioia approach, from selecting a research question to developing a data structure and grounded process model. Various techniques are covered, including a general *overview* of R programming language; the *installation* of R, RStudio, and RQDA; *data preparation, data coding techniques, data attributes, and "memo-ing" in RQDA; code abstraction* and *code plotting* in RQDA; and *grounded theory development* as the final output in qualitative research.

Our book is aimed at helping researchers – as well as undergraduate, Master, and Ph.D. to postdoctoral students – to *publish* their work in scholarly journals. Publishing in scholarly journals is a *tough* "game" in a researcher's life. This game is even tougher because of the relatively lower rate of acceptance of qualitative papers in top social science journals. Our logic in this book goes like this: if one has learnt the highest standard of qualitative scholarship (that we hope we have imparted to readers through this book), it will be relatively easier to apply it in the contexts that do not subject a research to double-blind, multiple-round review processes such as those in the industry, NGO, or policy-making fields. We believe that our book helps professionalize the qualitative craftsmanship in these non-scholarly settings.

We strongly encourage academic readers to start reading from the "author's reflection" above and then read the philosophical treatment on qualitative research (Chap. 1) and onwards. Because some of the chapters (e.g., Chap. 1) might sound a little too philosophical or "dry" for nonacademic readers, simply scan that chapter and quickly move to the core of the book via Chap. 3 and onwards. For experienced qualitative scholars who primarily use a "manual" approach in analyzing qualitative data, the best part to start is the coding process using RQDA (Chap. 8) right through to Chap. 11.

Overview of Chapters

In Chap. 1, we discuss the landscape of the qualitative research enterprise and articulate the important terminologies (linguistic devices) commonly used in qualitative research. We then discuss five different characteristics and "must-have" elements in qualitative research. Finally, we discuss three main paradigms in qualitative research, particularly the interpretivist-constructivist Gioia, the empiricist-positivist Yin-Eisenhardt, and the critical realist Langley approaches to qualitative research.

In Chap. 2, we discuss the advantages and caveats in using computers to assist with qualitative research (CAQDAS) and how the use of computers can increase transparency, trustworthiness, and rigor of qualitative research as opposed to manual analysis. We then review major CAQDAS tools along their technical specification, capabilities, and methodological strengths and how they fared with RQDA, the *R package for Qualitative Data Analysis,* software of our interest. We briefly discuss one of the most important functions for CAQDAS – *data coding, retrieval*, and *sharing* and leveraging actual examples of published research to demonstrate examples of data coding.

In Chap. 3, we present a six-step process to conduct CAQDAS-based qualitative research, from articulating the research question and conducting an initial literature review to discussing sampling approaches, crafting an interview protocol, performing data collection, and doing data analysis including creating a data structure and grounded theory development. These procedures should not be seen as a *formula* but rather as a guidance and inspiration for qualitative research. Essentially, all research follows more or less the steps described above. To illustrate the use of this approach, we review a study by the first author (Chandra 2017) on how a handful of ex-terrorists escape their brotherhood of extremists to pursue fast-food restaurant work – work integrated within mainstream society – and then transition into social entrepreneurship running such fast-food enterprises that employ others with similar backgrounds.

Chapter 4 reviews R programming language as a statistical computing platform and introduces RQDA as an extension (or package) of R geared for qualitative research. We discuss the merits of R and RQDA as open-source tools to support qualitative research. We also provide some examples of how to use *syntax* (a string of commands to instruct the software to execute certain tasks) to operate R and RQDA. We describe the RQDA console and the various functions available in RQDA for new project creation, data import, data export, etc.

In Chap. 5, we discuss the technical details for installing RQDA using R syntax in different operating systems, from Windows and Mac to Linux. We also demonstrate RQDA installation without using R syntax.

In Chap. 6, we demonstrate the techniques for data preparation in qualitative research. We show how to convert qualitative data into a format that is supported by RQDA, and techniques to combine several textual data files (e.g., interview transcripts or news articles) into a case file, and other data preparation tricks. Using a small qualitative study designed for illustrative purposes (the profiles of strategies and outcomes of social innovation drawn from a large social innovation database), we show how to conduct sampling and data preparation in a systematic manner.

In Chap. 7, we demonstrate the techniques to start and launch RQDA in an easy manner. We also show how to create a new project and manage a qualitative research project using RQDA. Lastly, we show how to import qualitative data files into RQDA.

In Chap. 8, we discuss and differentiate the *inductive* coding process in qualitative research and how to do inductive coding in RQDA. Using our sample project "*Social Innovation Strategies and Impacts in the Healthcare Sector*" as an illustration, we demonstrate ways to conduct data coding to create first-order codes/concepts as the first step in the Gioia approach.

In Chap. 9, we discuss various techniques to manage (textual) data files by assigning different attributes, creating memos, or aggregating files into larger case files to support qualitative data analysis in RQDA.

In Chap. 10, we demonstrate the techniques to abstract or aggregate first-order codes or concepts to the second-order themes in RQDA, to plot the codes in a two-dimensional network format, and to export the plotted codes in an HTML file for sharing with research collaborators and journal gatekeepers.

In Chap. 11, we show how to transform the coded data into a "data structure." Data structure is an aggregation from first-order codes (codes that capture meanings at the informant's level) to second-order themes (an abstraction of two or more first-order codes) and finally to aggregate theoretical dimensions (theoretical concepts that link data and theory) and how to reconstruct the static movie (data structure) into a *grounded process model* (a motion picture) – the key goal of qualitative research.

Chapter 12 offers our reflection on the advantages and shortfalls of the RQDA-based Gioia approach for qualitative research. We suggest future avenues to improve qualitative research using RQDA, to extend the use of R to support qualitative research and to conduct mixed-method research using R.

Contents

About the Authors

Yanto Chandra is Associate Professor in the Department of Applied Social Sciences, Faculty of Health and Social Sciences, at The Hong Kong Polytechnic University (PolyU). Chandra's research focuses on organizations that create hybrid (economic and social) value, sustainable development, and new forms of governance that increase societal welfare. His research is published in leading journals including the *Journal of Business Venturing*, *Journal of International Business Studies*, *VOLUNTAS, and World Development,* among others. He is an Associate Editor of the *Business Ethics: A European Review*, *Journal of Social Entrepreneurship* and also serves on the Editorial Board of *Social Enterprise Journal*. Among other awards, Dr. Chandra was the recipient of the Outstanding Paper Award 2010 by Emerald Publisher and the Winner of the City University of Hong Kong's Teaching Excellence Award 2016 and Teaching Innovation Award 2017. Prior to coming to PolyU, he was Associate Professor in the Department of Public Policy at the City University of Hong Kong (CityU) and Assistant Professor in the Marketing and Strategy Department at the University of Leeds (United Kingdom) and University of Amsterdam (Netherlands). Prior to these, he spent nearly 7 years in the corporate world in Hong Kong, Singapore, and Jakarta.

Shang Liang is a Ph.D. candidate in the Department of Public Policy at the City University of Hong Kong. She completed a master's degree in Public Policy and Management at the City University of Hong Kong and a bachelor's degree in Social Policy and Administration at the Hong Kong Polytechnic University. Her research focuses on social entrepreneurship, nonprofit organizations, and civil society. She is an expert in RQDA and R for both qualitative and quantitative research.

Chapter 1
Qualitative Research: An Overview

Abstract Qualitative research is one of the most commonly used types of research and methodology in the social sciences. Unfortunately, qualitative research is commonly misunderstood. In this chapter, we describe and explain the misconceptions surrounding qualitative research enterprise, why researchers need to care about when using qualitative research, the characteristics of qualitative research, and review the paradigms in qualitative research.

Keywords Qualitative research · Gioia approach · Yin-Eisenhardt approach · Langley approach · Interpretivism · Positivism

1.1 Why You Should Care About Qualitative Research

Qualitative research[1]—research that primarily or exclusively uses non-numerical data—is one of the most commonly used types of research and methodology in the social sciences. Unfortunately, qualitative research is commonly misunderstood. It is often considered "easy to do" (thus anyone can do it with no training), an "anything goes approach" (lacks rigor, validity and quality standards[2]), and is

[1] Qualitative research is defined as the practice used to study things — individuals and organizations and their reasons, opinions, and motivations, beliefs in their natural settings. It involves an observer (a researcher) who is located *in the field*, who transforms the world into a series of representations such as fieldnotes, interviews, conversations, photographs, recordings and memos (Denzin and Lincoln 2011). Many researchers employ qualitative research for exploratory purpose while others use it for 'quasi' theory testing approach. Qualitative research is a broad umbrella of research methodologies that encompasses grounded theory (Glaser and Strauss 2017; Strauss and Corbin 1990), case study (Flyvbjerg 2006; Yin 2003), phenomenology (Sanders 1982), discourse analysis (Fairclough 2003; Wodak and Meyer 2009), ethnography (Geertz 1973; Garfinkel 1967), and netnography (Kozinets 2002), among others. Qualitative research is often synonymous with 'case study research' because 'case study' primarily uses (but not always) qualitative data.

[2] The quality standards or evaluation criteria of qualitative research comprises: (1) *credibility* (that a researcher can provide confidence in his/her findings), (2) *transferability* (that results are more plausible when transported to a highly similar contexts), (3) *dependability* (that errors have been

Y. Chandra, L. Shang, *Qualitative Research Using R: A Systematic Approach*,
https://doi.org/10.1007/978-981-13-3170-1_1

"outdated" because it relies on manual techniques that have not changed since 50 years ago. These are misperceptions. Qualitative research is challenging to execute and defend, it needs to follow certain generally accepted principles and assumptions and can embrace the latest advances in computing technologies. Like a focused infrared laser, qualitative research can powerfully get to the heart of a social issue, belief, or value if the user is trained in using the infrared laser.

As we argue in this book, qualitative research is a challenging methodology to use, and it deserves respect from scholars from any epistemological background—from positivist, interpretivist, and simulation, to mathematical. The challenge is even taller if one is aiming high: to publish qualitative research in a top-tier scholarly journal. We also argue that qualitative research, when performed optimally and using valid approach(es), can achieve a high level of rigor, quality, and trustworthiness as all other methodologies.

Because the data in qualitative research are predominantly "words" (from interviews, field notes, conversations, recordings, and memos) as opposed to "numbers", qualitative researchers often face enormous pressure to *convince* readers and journal editors and reviewers (gatekeepers) that the conclusions are *plausible* and *defensible*. Qualitative researchers do not enjoy the privileges of "hiding behind the *p* values or statistical coefficients" as do quantitative researchers. The underlying problem is that there is little or no *consensus* among qualitative scholars regarding what high-quality qualitative research looks like. Another problem is that qualitative research is *tacit* in many ways. That is, researchers' decisions and actions during the data collection and analysis often remain hidden to outsiders. As some scholars have argued, qualitative research suffers from the "no accepted boilerplate" (no standardized approach) problem (Pratt 2009, p. 856). This assertion is partly true. There are legitimate paradigms in qualitative research that one can use but they are not obvious to many qualitative researchers. This has hampered the broader acceptance of the qualitative research enterprise.

To address these often-misplaced criticisms against qualitative research, we encourage qualitative researchers to couch their work using what scholars call *valid* approaches to qualitative work. We do this by demonstrating and extending the use of the *Gioia* approach, one of the valid methodologies for doing qualitative research. The Gioia approach is essentially about *theory building*[3] or *discovery* and seeks to generate and develop new *concepts* and *theories*[4]. In doing so, we take advantage of

minimized, proper documentation is provided), and (4) *confirmability* (that conclusions are internally consistent and supported by data) (see Lincoln and Guba 1985).

[3] We classify research into a continuum of *theory building— > theory elaboration— > theory testing*. Theory building is also known as theory exploration. Theory elaboration refers to the use of qualitative data and a method to seek "confirmation" of the relationships among variables or processes or mechanisms of a social reality (Bartunek and Rynes 2015).

[4] In the context of qualitative research, theory/ies usually refer(s) to conceptual model(s) or framework(s) that explain the relationships among a set of variables or processes that explain a social phenomenon. Theory or theories could also refer to general ideas or frameworks (e.g., institutional theory, emancipation theory, or identity theory) that are reviewed as background knowledge prior to the commencement of a qualitative research project.

the latest advances in computing technologies, in the form of an open source computing platform that runs in R called RQDA, to facilitate data analysis, storage and retrieval, and data presentation—thus seeking to professionalize qualitative research methodology and put it on par with other methodologies in terms of the use of advanced technology.

1.2 The Characteristics of Qualitative Research

Qualitative research is a 'big tent' that encompasses various schools of thoughts. There is a general consensus that qualitative research is best used to answer *why* and *how* research questions, but not *how much* or *to what extent* questions. The word 'how can[5]' is also frequently used in the research question of a qualitative research; this typically requires open-ended vs. closed-ended questions and answers (e.g., yes or no, ordinal or metric scales) answers. A researcher trained in a particular methodology, says laboratory experiments or econometrics, may see the world through a 'to what extent?' lens, or is only interested in solving problems dealing with a 'to what extent' problem. Likewise, a scholar trained in a qualitative methodology may see the world from a 'why and how' lens. Understanding this kind of orientation, or bias, is useful. Importantly, it is important to assess[6] a qualitative work based on its ontological and epistemological foundation (e.g., judging a qualitative work based on an interpretive-constructivist epistemology using an interpretive-constructivist methodology versus a positivist or critical realist methodology).

Before demonstrating the application of a systematic qualitative research a la the Gioia approach using the R platform, we discuss five general characteristics of qualitative research that we view as necessary to conduct qualitative research.

First, qualitative research is an *iterative* and *emergent* process. This means that qualitative research rarely relies on a one-off process in producing a finding. Rather, it relies on multiple back and forth comparisons between data and theory, and, quite frequently, among the analysis produced by data analysts involved in a research project. It is emergent because a qualitative researcher does not predetermine a research's outcome at the outset nor limit his/her observations and analysis based upon a certain preconception or idea, model, framework or theory at the beginning of a study. Rather, the researcher follows the logic of "letting the data speak" (allow the data to reveal a pattern or process) for itself. As a consequence, qualitative research often looks messier than quantitative research because the process is not linear.

Second, qualitative research is primarily employed to explore and *discover* a new understanding and or explanation, and/or to produce *new*, *surprising*, or *interesting*

---

[5] For example, a qualitative research can ask the following question: "How can institutional change succeed in social contexts that are dominated by organized crime?" (Vaccaro and Palazzo 2015).

[6] We have witnessed numerous cases in which committed positivist methodologists were asked to review qualitative papers, and they used a survey approach to assess the quality of an interpretivist work. This *reviewers' fallacy* is dangerous and hampers the progress of a field of research. Editors must be cognizant of such fallacy and avoid it.

findings. There is no point to conducting a qualitative study if the findings merely describe what has already been known or reported (i.e., purely *descriptive* research), such as reporting that, for instance, "social impact is important for social enterprises"[7], or fail to dig deeply beyond the obvious (i.e., *superficial* research), such as claiming that "social opportunity recognition precedes the development of social enterprises", or is simply used to verify[8] the relationships among variables (i.e., qualitative research that is quantitative in nature), such as "we found that A causes B in 97% of the cases". Thus, qualitative research is put to good use if it produces *new concepts* (e.g., concepts not previously mentioned in the literature), *counter-intuitive* or *surprising* findings (e.g., reporting that 'innovativeness weakens performance through mechanisms A, B and C'), or challenges conventional wisdom as explained by theory X (e.g., 'resource bricolage' can eventually speed up failure among new organizations in autocratic countries' with explanation of the underlying processes), or *interesting findings* (e.g., exploring previously unexplored phenomenon, or interview unique informants or contexts), such as finding that 'street demonstrations in new democracies in the developing world are not part of democracy if they are 'paid demonstrations' involving unemployed workers who make street protests for a living' and explain the why and how behind them, or any combination thereof.

In discovering a new concept or process, it is common for qualitative researchers to build or invent *new vocabularies* that best represent new concept(s) of interest. For instance, coining a new terminology that represents an overlooked social phenomenon (e.g., *sense making*, a new concept that explains leaders' role in instigating strategic change process (Gioia and Chittipeddi 1991)) or re-conceptualizing an existing phenomenon using a new perspective following insights that emerge from qualitative data (e.g., *"social entrepreneurship as an instrument of emancipation for ex-terrorists";* Chandra 2017a, b, p. 658).

Third, qualitative research typically relies on *rich* (layered, intricate, detailed, concrete, and nuanced) and *thick* data (in-depth data through multiple informants or a variety of sources). Qualitative research contains rich and thick description and or explanation of a social phenomenon (e.g., why and how ex-terrorists perceive

---

[7] A social enterprises (SE) is an organization that combines social welfare and commercial logics (Doherty et al. 2014), or that uses business principles to address social problems (Mair and Marti 2006); thus, qualitative research that reports that 'social impact' is important for SEs is too descriptive and, arguably, tautological. It is not uncommon to see authors submitting purely descriptive papers to scholarly journals.

[8] Some qualitative researchers have conducted qualitative work using primarily a checklist (ticking the boxes) to show the presence or absence of variables, as if it were a survey-based study. This is utterly inappropriate for a qualitative work. A qualitative work needs to show the richness and depth of qualitative findings. Nevertheless, it is acceptable to use such checklists as *supplementary* data if a study involves too many informants or variables of interest, or the data is too complex due to its longitudinal nature (e.g., a study that involves 15 cases observed and involving 59 interviews with 33 informants within a 7-year fieldwork used an excel sheet to tabulate the number of events that occurred as supplementary data to the main analysis; see Chandra 2017a, b).

the meaning of work in the post-prison life). During the data collection process[9], a researcher should make the best attempt to collect as much and as relevant (textual, visual) data as possible about the subject of interest such as conducting multiple in-depth interviews with an informant or an interview with a group of informants, for a more holistic and defendable data, and then add this data to other sources of data (e.g., news articles, observations, documents, website materials) on the same topic—known as *data triangulation.* This is the opposite from the conventional positivist epistemology (e.g., survey or laboratory experiment methodology), which tends to collect 'lean data' via a few questions and precise measurement scales (e.g., 1–5 scale).

Fourth, qualitative research seeks to produce *processes, theories,* and *progressions* that articulate and *explain*[10] a social phenomenon. To achieve these, a qualitative researcher will follow a methodology[11] (e.g., Gioia, or Eisenhardt, or Ragin approach) to analyze the qualitative data and abstract them to produce higher-level meanings, and map how one event (or process or concept) leads to other events (or processes or concepts). Therefore, qualitative research focuses on the *dynamics* of a social phenomenon ("how change unfolds") and seeks to study observed or recorded events to understand and explain *why* and *how* they produce outcomes. The insights generated by qualitative research is called *analytical generalization*[12] as opposed to statistical generalizations, which are insights generated by quantitative research.

Lastly, qualitative research is *contextual.* Although all research must consider the context, there is a general consensus that *context is king* in qualitative research. This means that first, identifying a *new or fresh context* that contributes[13] to our understanding of a field is an advantage. By doing so, qualitative research can offer *theory*

[9]As mentioned earlier, there are different types of qualitative research. Thus, a qualitative researcher will customize the data collection process to fit the type of research being conducted. For example, for researchers using ethnography, the primary data will be in the form of photos and/or videos and interviews; for those using netnography, the primary data will be internet-based textual data. Interview data is perhaps the most common type of data used across all types of qualitative research designs and is often synonymous with qualitative research.

[10]The purpose of qualitative research is to provide an *explanation*, not merely a *description* and certainly not a *prediction* (which is the realm of quantitative research). However, description is needed to illustrate qualitative data collected, and usually researchers describe their qualitative data by inserting a number of important "informant quotes" in the body of a qualitative research report.

[11]We advise qualitative researchers to adhere to one approach to avoid any epistemological and ontological mismatch that may arise among different camps in qualitative research. For instance, mixing a positivist with a constructivist approach in qualitative research frequently leads to unnecessary criticism and even rejection from journal editors and reviewers; it shows a lack of methodological competence or awareness of one's epistemological position.

[12]*Analytical generalization* is not generalization to some defined population that has been sampled, but to a "theory" of the phenomenon being studied, a theory that may have much wider applicability than the particular case studied (Yin 2003).

[13]There are different types of contributions. Typically, a researcher is expected to clearly articulate the *theoretical contributions* for a qualitative work submitted to a scholarly journal. Other types of contributions are *practical* (or *managerial*), common for business/management journals, and *policy*, common for policy related journals.

extension, by studying a phenomenon in a new context (e.g., how "fake news" is designed by activists for political purposes using WhatsApp as a new context of study). Second, context can be used as a *dimension* or *analytical tool* to help understand a social phenomenon in comparative contexts (e.g., to understand the motivations of social entrepreneurship, one can examine how the behavior and motivation of social entrepreneurs might differ from business entrepreneurs). Importantly, a context-sensitive research enables qualitative researchers to make stronger inferences within a context (achieving internal validity) before gradually studying other contexts in a comparative sense to enhance the inferences.

1.3 Major Paradigms in Qualitative Research

As mentioned above, qualitative research suffers from "a lack of boilerplate[14]" problem. In other words, it lacks a standard approach. Although some qualitative scholars resist any dominant design in qualitative research in support of methodological pluralism (e.g., Friese 2011; Cornelissen 2017; Welch et al. 2011), we have witnessed a growing tendency for scholarly communities to adopt and legitimize major qualitative research paradigms in recent years. The most identifiable paradigms are *interpretivism, positivism,* and *critical realism* (see Table 1.1). Among the different qualitative research approaches, the *Gioia* approach is essentially interpretivist-constructivist or what we label *systematic interpretivism.* That is, it relies on interpretivist epistemology but executed using a systematic approach. The *Yin-Eisenhardt* approach is empiricist-positivist or what we label as *qualitative positivism*; that is it applies positivist epistemology to qualitative data. The Langley approach is critical realist, or what we label as *pluralistic realism.* This means that it relies on critical realist epistemology but adopts a pluralistic ("no particular style is preferred") approach to doing research.

We present a comparison of the Gioia, Yin-Eisenhardt, and Langley qualitative approaches and describe their differences across 15 dimensions (see Table 1.1). We argue that these approaches provide *guidance* and *inspiration* for qualitative scholars rather than to limit or *constrain* them. There is plenty of room for creativity and improvisation in applying each of these approaches. Also, these approaches are not 'the final word' on how qualitative research should be performed but instead are constantly evolving and improving. In essence they serve as maps to help guide scholars across the vast terrain of qualitative scholarship.

[14] There is ongoing debate on whether a template for qualitative research is desirable or necessary, with one camp of scholars (the pluralistic critical realists) that advocates a pluralistic approaches to qualitative research ("qualitative research should not follow a particular template or be prescriptive in its process") and the other camps are advocating for some form of consensus via the use of particular approaches (e.g., the Eisenhardt or Gioia Approach, etc.). However, as shown in Table 1.1, even the pluralistic critical realism in itself is a template and advocates an alternative form of consensus through the use of diverse and pluralistic approaches in doing qualitative research.

Table 1.1 Three approaches in qualitative research: a comparison

Dimensions	Gioia approach (Systematic interpretivism)	Yin-Eisenhardt (YE) approach (Qualitative positivism)	Langley approach (Pluralistic realism)
Epistemology of (or ways of understanding) social reality	*Interpretivism-Constructivism*: Social reality is interpreted and facts and events are made sense of—collectively constructed and understood.	*Empiricism/Falsificationism-Positivism*: Social reality exists 'out there' waiting to be found, described, explained, predicted and understood.	*Critical realism*: Social reality exists independently of human observers; comprises (1) 'real world' structures and events that create social reality, (2) 'actual world' events and structures that occurred regardless of whether or not we observe, and (3) 'the empirical world', events that we can experience and/or measure. Embraces both *interpretivism* and *positivism*.
Style of theorizing	Concept-oriented process theorizing: Produces concepts that portray and explain the dynamics of social reality.	Variable-oriented theorizing: Produces causal laws that predict relationships among variables regardless of context.	Event-oriented process theorizing: traces causal processes that explain how and why things emerge, develop, and terminate in specific contexts. *Build typologies* that explain different patterns of change.
Objective	Theory building, theory exploration, theory discovery. Discovering new concepts, creating new vocabularies to explain social reality.	Theory testing/falsifying, theory refinement, theory replication, and elimination of alternative explanation(s).	Develop theory inductively and deductively (abductive process, or 'retroduction') to reconstruct, re-describe, and re-contextualize social reality.

(continued)

Table 1.1 (continued)

Dimensions	Gioia approach (Systematic interpretivism)	Yin-Eisenhardt (YE) approach (Qualitative positivism)	Langley approach (Pluralistic realism)
Key references and research logic	The grounded theory approach of Strauss and Corbin (1990), Glaser and Strauss (2017), Gioia et al. (2013) (i.e., from codes, to concepts to categories to building "theory").	Yin's (2003) case experiment logic and Eisenhardt's (1989a, b) case replication logic: one case is (dis)confirmed by other cases to reach theoretical saturation.	Langley et al.'s (2013) process theorizing: cross-case replication, longitudinal replication, event mapping, analytical generalization, and contextualized explanation.
The role of research questions, constructs, propositions, theory	Research question is developed at the start of study; a priori constructs do not prevail, nor do propositions, or theory.	Research question, a priori constructs, or propositions or "theory" are developed at the start of a study.	Research question is developed at the start of a study; theory does not dictate the direction of a research.
Sampling logic	*Purposive sampling* (e.g., extreme or deviant cases; rare revelatory cases that offer new insights); also uses *theoretical sampling*, with data collection largely guided by theoretical insights emerging from earlier phases of data collection.	*Theoretical sampling*: cases are chosen for theoretical reasons such as replication (e.g., high vs. low replication), extension of theory, contrary replication, and elimination of alternative explanations.	*Context variation sampling:* multiple cases chosen to study a phenomenon; uses *constrained variation sampling*: cases selected from the same initial sample but follow different pathways that lead to different outcomes.
Size of samples	*Single case* and *multiple cases* (up to 4 cases, occasionally more) aimed to explore and explain rich, interesting cases (vs. theoretical saturation)	Preference for *multiple cases* (6–10 cases or more) aimed to confirm/disconfirm/falsify a 'theory' of interest.	Preference for *multiple cases* (2–3 longitudinal cases) with a relatively large number of interviews (29–100 interviews) to explain change processes and develop typologies that classify events into different meanings.
Nature of data collected	Primarily *qualitative data* (e.g., in-depth interviews collected in multiple rounds); also *supplementary data* (e.g., news articles, website information, organizational documents).	*Qualitative* data (e.g., in-depth interviews, observations, secondary data) and *quantitative* data (e.g., questionnaires, checklists of variables, financial data, health data).	Primarily *qualitative* data (e.g., in-depth interviews, observations, secondary data, archival data).

Role of interview protocol	To flexibly adjust the interview protocol as the data collection progresses; perform follow-up interview with informants based on insights from subsequent interviews.	Need a well-developed interview protocol to measure and capture social reality in a precise manner.	To flexibly adjust the interview protocol as the data collection progresses so as to explore multiple explanations.
Research process and intended final output	Develop a *data structure* to represent the overall data coded by assembling first-order concepts (via open coding), second-order themes (via axial coding), and aggregate dimensions; leading to a *grounded process model* and new concept development.	Conduct within-case analysis followed by cross-case analysis; then *tabulate evidence* to refine constructs and identify *testable propositions*.	Map the events and temporal patterns of multiple cases; then develop event sequence of each case in an event map that illustrates the mechanisms of social reality.
Presentation of research findings	No fixed style. Usually starts with a *data structure* and *process model* (concepts and their relationships), then present proof quotes as *data tables* and power quotes as *indented paragraphs* in the report	Start with *propositions or theoretical framework*, then present *tables of evidence* (with short quotes, some numerical data, other constructs observed), finally a *box and arrows* model	No fixed style. Uses particular event mapping (to unpack particular cases) and a general event mapping.
Analytical process	Primarily use *manual* data analysis, but a few researchers use CAQDAS such as NVivo	Primarily use *manual* data analysis (using pens/pencils, paper, highlighter, folders, etc.)	No specific process. Uses process tracing (constructing a causal link from outcome backward to events); relies on abductive process leading to re-description or re-contextualization

(continued)

Table 1.1 (continued)

Dimensions	Gioia approach (Systematic interpretivism)	Yin-Eisenhardt (YE) approach (Qualitative positivism)	Langley approach (Pluralistic realism)
Key phrases	Creativity with fences	Replication logic, pattern matching, time-series analysis	Best to use several methodologies. Several possible explanations of an event or outcome
	Data structure serves as evidence	Produce generalizable and testable theory	No best way to perform process research
	Avoids "cherry picking" data	Escape the idiosyncratic detail of individual cases to achieve context-free propositions	Studying phenomena over time
	Data structure is a static picture; the grounded model converts that picture into a dynamic movie		Visual mapping; temporal bracketing
Examples of important findings	The discovery of "sense giving", "identity ambiguity", "transitional identity" concepts (Gioia et al. 2013)	A quick decision-making model (Eisenhardt 1989a, b); consumers' material possessions as irreplaceable or replaceable (Grayson and Shulman 2000)	Reflective and experimental "spaces" in change processes in large organizations (Bucher and Langley 2016); a typology of identity work in merging organizations ("mavericks", "fighters", "adapters", and "victims"; Langley et al. 2012)
Key papers	Methods: Strauss and Corbin (1990), Gioia et al. (2013) Empirical: Gioia et al. (2010), Vaccaro and Palazzo (2015), Chandra (2017a, b).	Methodological: Yin (2003), Yin (1981), Eisenhardt (1989a) Empirical: Eisenhardt (1989b), Graebner and Eisenhardt (2004).	Methodological: Langley et al. (2013), Langley and Abdallah (2011) Empirical: Denis et al. (2001), Langley et al. (2012), Bucher and Langley (2016)

Interpretivism is a school of thought that treats social reality as subjective and co-constructed by researchers aiming to make sense of the world through individual or collective experiences (Gioia 2007; Ramoglou and Tsang 2016; Stake 2010; Young and Collin 2004). In this school, *language* and linguistic devices convey a collectively constructed and understood social reality. In this school, the role of qualitative methods is to interpret, explore, and discover new concepts, constructs, working theories or models or frameworks. Qualitative research is rooted in interpretivism. Interpretivism itself is a 'big tent' that encompasses symbolic interactionism, hermeneutics, dramaturgy, ethnomethodology, and ethnography, among other fields (for more, see Prasad 2018).

Positivism, on the other hand, is a school that treats social reality as objective, something that "exists out there" for verification and testing (Brinkmann 2018; Hunt 1991; King et al. 1994). This school is inspired by the hypothetico-deductive paradigm in the social sciences. It uses quantitative methods but also supports the application of qualitative methods to elaborate or validate preconceived constructs, "theories" or propositions (Dubé and Paré 2003; Langley and Abdallah 2011).

Interpretivism is often used for *theory-building* and *discovery-oriented* research, while positivism for *theory-testing or theory-elaboration* or *theory-refinement* research. Depending on one's epistemological position, some scholars argue that interpretivism and positivism are complementary (Lin 1998; Sobh and Perry 2006) because social reality is partly real, partly co-constructed and interpreted and hence structured in various levels; thus both camps can be combined to better study the social reality. This gave rise to the *critical realism* school (Easton 2010; Ramoglou and Tsang 2016; Welch et al. 2011). In relation to the other two schools, critical realism is *pluralistic* in its approach (Gehman et al. 2017) because, in the words of Langley, "a single [qualitative] methodology is really not a good idea" (p. 6) and "it is better to welcome [methodological] diversity" (p. 10).

The Gioia Approach

The Gioia approach (Gioia et al. 2013) is inspired by the interpretivist-constructivist epistemology and is more concerned with obtaining insight and generating, exploring and discovering new concepts, processes, or mechanisms, and finally generating a process model or a "theory". The Gioia approach uses a systematic and rigorous approach to doing qualitative research as a response to the "impressionistic" (and often "poetic") and "insufficient demonstrative evidence" (Gioia 2019, forthcoming) criticisms on qualitative research. The Gioia approach has an evidentiary basis and storytelling component. It shows and tells readers how and why a researcher arrives at his/her assertions based on the data analyzed. This approach conceptualizes qualitative research as a sense-making process in that a researcher relies on sense-making processes to figure out what he/she is confronting.

The Gioia approach typically begins with a research question, however, it frequently includes a basic discussion of the relevant literature regarding a phenomenon of interest (e.g., Gioia et al. 2010; Vaccaro and Palazzo 2015) before presenting the methodology section. This structure clarifies definitions, acknowledges the knowns and unknowns in a particular field, and/or highlights inconsistencies and acknowledges the absence of theoretical explanations for a phenomenon. In this approach, a researcher is not dictated by the initial literature review. In terms of data collection, the Gioia approach generally uses interview data, embraces data triangulation, and includes longitudinal data collection and multiple informants. This approach primarily relies on *purposeful sampling* (Lincoln and Guba 1985) but does not recommend a set number of case sites (as opposed to the Yin-Eisenhardt approach, see below) because it is more concerned with the robust interpretation of themes emerging from the data than with replication logic (as in positivism). Thus, the Gioia approach has been used in single-case studies (e.g., Corley and Gioia 2004; Nag et al. 2007), although we argue that it is equally valid when applied to multiple-case studies involving four or more cases.

The Gioia approach's uniqueness is in its process of moving back and forth between the emergent findings and the literature/theories to identify, generate or position new concepts in or from the data collected (Alvesson and Kärreman 2007). In so doing, it relies on a unique process of coding the (textual) data as *first-order categories* (called "open coding") with labels that represent the raw data as closely as possible (e.g., "our forests are disappearing" and "the rivers are polluted" can be coded as "environmental destruction" as a first-order code) and then abstracting them into *second-order themes* within a firmer theoretical realm (called "axial coding") (e.g., first order codes "environmental destruction" and "many species become extinct" can be abstracted into "dying ecosystem" as a second-order theme) and, finally, into *aggregate dimensions* (called "selective coding) (e.g., second order themes "dying ecosystem", "other habitable planets exist" and "it's time to get out of the Earth" can be abstracted into "new habitable planets as a solution" as the aggregate dimensions). This process is facilitated by one member of the research team acting as 'devil's advocate' to critique the interpretations that emerge (Gioia et al. 2013). The three orders of data analysis are structured and portrayed as a *data structure*. In a nutshell, the data structure offers a bird's eye view of insights from the data and is an effective method of both communicating findings and enabling readers/reviewers to easily follow the researcher's reasoning. Data tables containing *representative data* of informants' quotes (or proof quotes) are typically included as evidence, whereas quotes are also presented in the manuscript (the power quotes) itself to highlight new concepts or insights. Finally, the Gioia approach aggregates the overall coding output into a *process model* or *grounded theory model* that depicts the dynamic relationships among the emergent concepts. In other words, it adopts a *concept-oriented process-based theorizing*; that is, it seeks to produce (static and dynamic) concepts that portray and explain the dynamics of social reality.

What Gioia understates is that the process of constructing the data structure and grounded theory model is messy and iterative. It is inherently a creative process but performed with a well-organized imagination (Weick 1989). Thus the Gioia approach can be described as "creativity with fences".

Yin-Eisenhardt Approach

In contrast, the Yin-Eisendhardt approach (Eisenhardt 1989a; Yin 2003) is rooted in positivist epistemology and primarily focuses on *theory or construct validation, elaboration, refinement, disconfirmation, or falsification*. It seeks to develop *theoretical propositions* as the ultimate product. This approach is also known as *variable-oriented theorizing*; whereby a researcher produces causal laws that predict relationships among variables or constructs regardless of context.

Pursuant to this approach, a researcher typically begins with a research question and some knowledge of the relevant literature regarding a phenomenon of interest but makes little effort to explicate the theory. Typically, the methodology is presented immediately after the Introduction, and little or no literature review is presented because this approach avoids being constrained by what is already known in the literature (e.g., Eisenhardt 1989a, which has no literature review; Ozcan and Eisenhardt 2009, which has only a brief literature review).

Data collection under the Yin-Eisendhardt approach typically requires data triangulation, multiple cases, multiple informants, and often involves some form of quantitative data. Through *replication logic* (Eisenhardt 1989a; Yin 2003), a researcher seeks to find similarity in the potential variables of interest from one case to the next until theoretical saturation is reached. To achieve this saturation, the researcher investigates multi-case sites where s/he interviews numerous informants (Eisenhardt 1989a; Ozcan and Eisenhardt 2009). For example, if laziness is theorized to improve work efficiency under highly stressful work environment', then a researcher will conduct interviews with managers in various companies to see if laziness eventually improves work efficiency. Eisenhardt (1989a, b) suggests between 6 and 10 cases for a solid qualitative study. The researcher generally begins to analyze each case alone (called 'within-case analysis') and then proceeds to conduct a cross-case analysis (Yin 2003). As patterns emerge, the researcher induces robust theoretical constructs and causal relations from the cases, following an iterative process of cycling among theory, data and the literature to refine the findings and clarify contributions. This approach typically *portrays* the data containing potential variables of interest in one or more tables and in tables containing key descriptive, explanatory, or illustrative quotes as evidence. This approach presents *theoretical propositions* as causal statements that resemble those often found at the beginning of a hypothetico-deductive paper, as the ultimate product.

A key feature of the Yin-Eisenhardt approach is how it wraps up a qualitative research project (or 'reaches closure'). This approach tends to focus on *tabulating evidence* to refine constructs and their relationships, by focusing on variables that influence a dependent variable (e.g., decision durations, decision types, stakeholders types that influence high vs. low firm performance (see Eisenhardt 1989a)), using a few *short informant quotes*, presenting *testable propositions,* and representing the overall findings in a *box-and-arrow model* to demonstrate causality.

The Langley Approach

The Langley approach is inspired by the critical realism school. This approach can be labeled as *pluralistic realism* because it advocates the use of diverse and pluralistic approaches in doing qualitative research including the interpretivism and positivism. This approach adopts an *event-oriented process-based theorizing*, wherein a researcher traces the causal processes that explain how and why things emerge, develop, and terminate in specific contexts. In doing so, it embraces inductive and deductive (or abductive process, or 'retroduction') processes in order to reconstruct, re-describe, and re-contextualize social reality. It does not, however, seek to invent new concepts, which is commonly used in the Gioia approach. However, akin to the Gioia approach it embraces a process-oriented style of theorizing in that it produces event maps that explain a dynamic social process.

The Langley approach embraces a series of tools to develop process theorizing: cross-case replication (similar to the Yin-Eisenhardt approach), longitudinal replication, event mapping, analytical generalization, and contextualized explanation. *Context* is seen as king; it allows the study of multiple pathways to explain a social reality. Tensions, dialectics, and equifinality (i.e., where the same initial state can lead to multiple outcomes) from the data are considered positive. Most research that adopts the Langley approach involves *multiple cases*, which are often much smaller than those suggested by Yin-Eisenhardt or Gioia—in the range of 2–3 case sites but with a large number of interviews (between 29 to 100 interviews) conducted over 2 or more years.

In the Langley approach, the research outputs typically comprise event maps and temporal explanations that explain the micro (specific cases) and macro (across cases) mechanisms of social reality (e.g., how organizations radically change routines, how merging organizations struggle to create identities for themselves and others). Although Langley claims that her approach does not constitute a specific process, methodology or consensus-seeking goal, she in fact does. She primarily advocates the use of *process tracing* (construction of a causal link from outcome to events in the form of *process maps*) and relies on the *abductive* process leading to re-description or recontextualization of a social phenomenon (e.g., reinterpreting and re-classifying the meaning of identity among merging organizations or the role of spaces in how organizations break routines).

A good example of the Gioia approach is a study of young social activists who successfully challenged the practice of paying protection money Italian criminal organizations (Vaccaro and Palazzo 2015). These scholars gradually coded various first-order concepts from primarily interview data conducted with the activists, local business owners, local citizens (consumers, students, teachers), to police and mafia experts, into six second-order themes (i.e., which comprise new concepts called values, hooking, anchoring, activating, securing, and uniting) and finally aggregating them into two major dimensions (i.e., moralizing and integrating). "Power quotes" (i.e., the most powerful or revelatory examples of informants' speeches or statements) are also often used in the text and data tables of Gioia type of studies. For example, Vaccaro and Palazzo (2015) creatively inserted short quotes of their informants in the text and used tables to summarize various important themes and the corresponding informants' quotes to portray the data and their abstractions (e.g., "*their strategy to group people together and to work with multiple networks is very successful in minimizing the risk of Mafia attacks*" to highlight a first-order code called "uniting", p.1088).

A typical example of the Yin-Eisenhardt approach can be seen in Eisenhardt (1989b) paper that ended the study by developing a model of strategic decision making in high-velocity environments along with propositions that theorized the relationships between on decision-making variables and firm performance. Eisenhardt (1989b) used qualitative and quantitative data to provide numerical estimates of the speed of decision making in the fast-changing technology sector (i.e., using decision durations as data, number of meetings, and classification of active versus passive conflict resolution to see variations among cases) and she used very limited informants' quotes in the data presentation.

A good example of Langley approach is the Bucher and Langley (2016) paper that provides a contextualized explanation of how organization members strived to radically change organizational routines – via the use of reflective and experimental spaces (spaces to imagine and spaces to integrate and act, respectively) and uses micro and macro event maps to portray the mechanisms in interruption and reorienting of routines in organizations. This paper did not offer any new concepts or confirm/disconfirm any variables but offered an event analysis across time that identified the unique steps of how routines are interrupted and reoriented.

The Gioia approach is a form of *systematic interpretivism* and is best used when a researcher focuses on meaning and concepts discovery and/or generation. It produces systematically derived concepts and their relationships that describe and explain a social phenomenon (e.g., sense-making and sense giving or moralizing as new concepts). The Yin-Eisenhardt approach is a form of *qualitative positivism* and is best used when a researcher seeks to focus on the confirmation/disconfirmation and testing of variables of interest. It produces variables and their relationships that are testable in the form of hypotheses. The Langley approach is a form of *pluralistic critical realism*—which encourages the use of diverse types of analytical techniques to unpack different layers of reality—and is best used when a researcher wants to understand time-sensitive events, and how events or processes or organizations evolve. It produces a variety of time-sensitive maps explaining how events unfold over time.

The choice as to which approach to use depends on one's research goal(s) and the type of data available (data is the most precious resource in any qualitative research). For instance, data representing events that are time-stamped might easily call for the use of the Langley approach. When researchers have access to numerical data such as an organization's performance data and interview data, then the Yin-Eisenhardt approach is feasible. When the data are primarily interview data (i.e., words), the Gioia approach is a natural choice. However, as we will describe later in Chap. 3 and beyond, the use of R platform enables the analysis of a large number of qualitative data (e.g., hundreds to tens of thousands of textual documents) and cases thus advancing the Gioia approach.

There is room for *creativity* and *improvisation* in qualitative research. We can choose between the Gioia, Yin-Eisenhardt or Langley approach, depending on the goal. Each of them can serve as an anchor to conduct high quality qualitative research. Having no methodological anchor at all (i.e., not knowing clearly which dominant approach is used in a qualitative study) weakens a qualitative research project and makes it difficult to defend qualitative findings in the journal review process.

These approaches are *constantly evolving* and should NOT be taken as the 'final word' or only approaches in qualitative research. For instance, the earlier version of Gioia's work (1992) that examined the Pinto Fires case used *descriptive interpretivism,* an approach advocated by Stake (2010) to describe social phenomena in a descriptive manner. And Gioia has continued to refine his approach. Likewise, Langley's (1988) research on the roles of strategic planning Welch et al.'s (2008) and other work in which she was not the main author typically employed the *descriptive interpretivist* approach. Over time these scholars developed more sophisticated approaches. In their recent work, Bucher and Langley 92,016) first presented an abstract form of a general process model (or what they called "first order findings") and subsequently offered concrete explanations and examples using the cases (or "second order findings). In this way, this looks like the reverse of Gioia's approach in that the Gioia's second order is more general and abstract that the first order analysis.

We argue that a software-assisted qualitative research approach can enhance the Gioia, Yin-Eisenhardt, and Langley approaches. In this book, we selected the Gioia approach to extend and enrich by harnessing R open-source software to enable a systematic, rigorous yet creative approach to qualitative research. Some scholars have criticized Gioia for "ruining the presentation of pure ethnography", employing a "naïve empiricism" approach, for being "too structured", "too stylized", "too inhabiting of creative insight" among other criticism (see Gioia 2019). If 'a way of seeing is a way of not seeing, then a way of not seeing is essentially a way seeing.' Thus, even methodological pluralists can be trapped by a particular view of theorizing (i.e., by not proposing a particular qualitative research approach). What matters more is one's awareness of the various dominant approaches to qualitative research and treating such approaches as guidance and inspiration for one's own qualitative research, rather than restrictions.

1.4 Conclusion

While seemingly easy, qualitative research is as challenging as quantitative research methodologies. Anyone can sing but to sing like opera stars Luciano Pavarotti or Renee Fleming requires sophisticated methodology and training. The same is true for qualitative research. Good advice for any qualitative research scholar is to learn the methodology, practice it and tweak it when necessary.

A major source of the difficulty in doing qualitative research lies in the tacit and non-standardized nature of this type of research. In the eyes of novice researchers, the differences between high quality and low-quality qualitative work is often imperceptible. In fact, differences can be observed. Examples include interesting and novel concepts or process model that challenges existing assumptions, or revealing a process that was taken for granted or offering a new perspective. The experience of being "grilled" by editors in top-tier journals is invaluable and often a better teacher than following a well-established step-by-step research formula. Sincerely hope that we have de-mystified qualitative research by exposing its common misconceptions, articulating its characteristics and requirements, and clarifying three types of approaches: the interpretivist-constructivist Gioia, the positivist Yin-Eisenhardt, and the critical realist Langley. Lastly, one should not evaluate a qualitative research approach (e.g., a Gioia approach) using the logic of another qualitative approach (e.g., Langley approach) for they are essentially different ontologically and epistemologically.

References

Alvesson, M., & Kärreman, D. (2007). Constructing mystery: Empirical matters in theory development. *Academy of Management Review, 32*(4), 1265–1281.

Bartunek, J. M., & Rynes, S. L. (2015). Qualitative research: It just keeps getting more interesting! In *Handbook of qualitative organizational research* (pp. 41–55). New York: Routledge.

Brinkmann, S. (2018). *Philosophies of qualitative research.* New York: Oxford University Press.

Bucher, S., & Langley, A. (2016). The interplay of reflective and experimental spaces in interrupting and reorienting routine dynamics. *Organization Science, 27*(3), 594–613.

Chandra, Y. (2017a). A time-based process model of international entrepreneurial opportunity evaluation. *Journal of International Business Studies, 48*(4), 423–451.

Chandra, Y. (2017b). Social entrepreneurship as emancipatory work. *Journal of Business Venturing, 32*(6), 657–673.

Corley, K. G., & Gioia, D. A. (2004). Identity ambiguity and change in the wake of a corporate spin-off. *Administrative Science Quarterly, 49*(2), 173–208.

Cornelissen, J. P. (2017). Preserving theoretical divergence in management research: Why the explanatory potential of qualitative research should be harnessed rather than suppressed. *Journal of Management Studies, 54*(3), 368–383.

Denis, J. L., Lamothe, L., & Langley, A. (2001). The dynamics of collective leadership and strategic change in pluralistic organizations. *Academy of Management Journal, 44*(4), 809–837.

Denzin, N. K., & Lincoln, Y. S. (2011). Introduction. In N. K. Denzin & Y. S. Lincoln (Eds.), *The Sage handbook of qualitative research* (4th ed.). Thousand Oaks: Sage.

Doherty, B., Haugh, H., & Lyon, F. (2014). Social enterprises as hybrid organizations: A review and research agenda. *International Journal of Management Reviews, 16*(4), 417–436.

Dubé, L., & Paré, G. (2003). Rigor in information systems positivist case research: Current practices, trends, and recommendations. *MIS Quarterly, 27*(4), 597–636.

Easton, G. (2010). Critical realism in case study research. *Industrial Marketing Management, 39*(1), 118–128.

Eisenhardt, K. M. (1989a). Building theories from case study research. *Academy of Management Review, 14*(4), 532–550.

Eisenhardt, K. M. (1989b). Making fast strategic decisions in high-velocity environments. *Academy of Management Journal, 32*(3), 543–576.

Fairclough, N. (2003). *Analysing discourse: Textual analysis for social research.* Abingdon: Routledge.

Flyvbjerg, B. (2006). Five misunderstandings about case-study research. *Qualitative Inquiry, 12*(2), 219–245.

Friese, S. (2011). Using ATLAS.ti for analyzing the financial crisis data [67 paragraphs]. *Forum Qualitative Sozialforschung/Forum: Qualitative Social Research, 12*(1), Art. 39. http://nbn-resolving.de/urn:nbn:de:0114-fqs1101397

Garfinkel, H. (1967). *Studies in ethnomethodology*. Malden: Blackwell Publishers.
Geertz, C. (1973). *Interpretation of cultures*. New York: Basic Books.
Gehman, J., Glaser, V. L., Eisenhardt, K. M., Gioia, D., Langley, A., & Corley, K. G. (2017). Finding theory–method fit: A comparison of three qualitative approaches to theory building. *Journal of Management Inquiry, 27*, 284–300. in press.
Gioia, D. A. (1992). Pinto fires and personal ethics: A script analysis of missed opportunities. *Journal of Business Ethics, 11*(5–6), 379–389.
Gioia, D. A. (2007). Individual epistemology – Interpretive wisdom. In E. H. Kessler & J. R. Bailey (Eds.), *The handbook of organizational and managerial wisdom* (pp. 277–294). Thousand Oaks: Sage.
Gioia, D. (2019). If I had a magic wand: Reflections on developing a systematic approach to qualitative research. In B. Boyd, R. Crook, J. Le, & A. Smith (Eds.), *Research methodology in strategy and management*. https://books.emeraldinsight.com/page/detail/Standing-on-the-Shoulders-of-Giants/?k=9781787563360
Gioia, D. A., & Chittipeddi, K. (1991). Sensemaking and sensegiving in strategic change initiation. *Strategic Management Journal, 12*(6), 433–448.
Gioia, D. A., Price, K. N., Hamilton, A. L., & Thomas, J. B. (2010). Forging an identity: An insider-outsider study of processes involved in the formation of organizational identity. *Administrative Science Quarterly, 55*(1), 1–46.
Gioia, D. A., Corley, K. G., & Hamilton, A. L. (2013). Seeking qualitative rigor in inductive research: Notes on the Gioia methodology. *Organizational Research Methods, 16*(1), 15–31.
Glaser, B. G., & Strauss, A. L. (2017). *Discovery of grounded theory: Strategies for qualitative research*. New York: Routledge.
Graebner, M. E., & Eisenhardt, K. M. (2004). The seller's side of the story: Acquisition as courtship and governance as syndicate in entrepreneurial firms. *Administrative Science Quarterly, 49*(3), 366–403.
Grayson, K., & Shulman, D. (2000). Indexicality and the verification function of irreplaceable possessions: A semiotic analysis. *Journal of Consumer Research, 27*(1), 17–30.
Hunt, S. D. (1991). Positivism and paradigm dominance in consumer research: Toward critical pluralism and rapprochement. *Journal of Consumer Research, 18*(1), 32–44.
King, G., Keohane, R. O., & Verba, S. (1994). *Designing social inquiry: Scientific inference in qualitative research*. Princeton: Princeton University Press.
Kozinets, R. V. (2002). The field behind the screen: Using netnography for marketing research in online communities. *Journal of Marketing Research, 39*(1), 61–72.
Langley, A. (1988). The roles of formal strategic planning. *Long Range Planning, 21*(3), 40–50.
Langley, A., & Abdallah, C. (2011). Templates and turns in qualitative studies of strategy and management. In *Building methodological bridges* (pp. 201–235). Bingley: Emerald Group Publishing Limited.
Langley, A., Golden-Biddle, K., Reay, T., Denis, J. L., Hébert, Y., Lamothe, L., & Gervais, J. (2012). Identity struggles in merging organizations: Renegotiating the sameness–difference dialectic. *The Journal of Applied Behavioral Science, 48*(2), 135–167.
Langley, A. N. N., Smallman, C., Tsoukas, H., & Van de Ven, A. H. (2013). Process studies of change in organization and management: Unveiling temporality, activity, and flow. *Academy of Management Journal, 56*(1), 1–13.
Lin, A. C. (1998). Bridging positivist and interpretivist approaches to qualitative methods. *Policy Studies Journal, 26*(1), 162–180.
Lincoln, Y. S., & Guba, E. G. (1985). *Naturalistic inquiry*. Beverly Hills: Sage.
Mair, J., & Marti, I. (2006). Social entrepreneurship research: A source of explanation, prediction, and delight. *Journal of World Business, 41*(1), 36–44.
Nag, R., Corley, K. G., & Gioia, D. A. (2007). The intersection of organizational identity, knowledge, and practice: Attempting strategic change via knowledge grafting. *Academy of Management Journal, 50*(4), 821–847.
Ozcan, P., & Eisenhardt, K. M. (2009). Origin of alliance portfolios: Entrepreneurs, network strategies, and firm performance. *Academy of Management Journal, 52*(2), 246–279.

Prasad, P. (2018). *Crafting qualitative research: Beyond positivist traditions*. New York: Taylor & Francis.

Pratt, M. G. (2009). From the editors: For the lack of a boilerplate: Tips on writing up (and reviewing) qualitative research. *Academy of Management Journal, 52*(5), 856–862.

Ramoglou, S., & Tsang, E. W. (2016). A realist perspective of entrepreneurship: Opportunities as propensities. *Academy of Management Review, 41*(3), 410–434.

Sanders, P. (1982). Phenomenology: A new way of viewing organizational research. *Academy of Management Review, 7*(3), 353–360.

Sobh, R., & Perry, C. (2006). Research design and data analysis in realism research. *European Journal of Marketing, 40*(11/12), 1194–1209.

Stake, R. E. (2010). *Qualitative research: Studying how things work*. New York: Guilford Press.

Strauss, A., & Corbin, J. M. (1990). *Basics of qualitative research: Grounded theory procedures and techniques*. Thousand Oaks: Sage.

Vaccaro, A., & Palazzo, G. (2015). Values against violence: Institutional change in societies dominated by organized crime. *Academy of Management Journal, 58*(4), 1075–1101.

Weick, K. E. (1989). Theory construction as disciplined imagination. *Academy of Management Review, 14*(4), 516–531.

Welch, C. L., Welch, D. E., & Hewerdine, L. (2008). Gender and export behaviour: Evidence from women-owned enterprises. *Journal of Business Ethics, 83*(1), 113–126.

Welch, C., Piekkari, R., Plakoyiannaki, E., & Paavilainen-Mäntymäki, E. (2011). Theorising from case studies: Towards a pluralist future for international business research. *Journal of International Business Studies, 42*(5), 740–762.

Wodak, R., & Meyer, M. (Eds.). (2009). *Methods for critical discourse analysis*. London: Sage.

Yin, R. K. (1981). Life histories of innovations: How new practices become routinized. *Public Administration Review, 41*, 21–28.

Yin, R. (2003). *Case study research: Design and methods*. Thousand Oaks: Sage.

Young, R. A., & Collin, A. (2004). Introduction: Constructivism and social constructionism in the career field. *Journal of Vocational Behavior, 64*(3), 373–388.

Chapter 2
Computer-Assisted Qualitative Research: An Overview

Abstract Computer-assisted qualitative data analysis (CAQDAS)-based qualitative research is both an *art* and a *science*. It relies on creativity and disciplined imagination (Weick KE, Acad Manag Rev 14(4):516–531, 1989) plus systematic, accurate, reliable, and iterative approaches. The primary challenge in qualitative research is to transform "hundreds of pages of field notes to a final report" (Miles MB, Huberman AM, Qualitative data analysis: an expanded source book. 2nd ed. Sage, Thousand Oaks, 1994, p. 281) in a rigorous, defendable, and auditable process. Qualitative research is not "a disorganized stumble through a mass of data, full of 'insightful' observations of a mainly anecdotal nature" (Silverman D, Interpreting qualitative data: methods for analysing talk, text, and interaction. Sage, London, 1993, p. 43). Rather, it is a disciplined yet creative way to bring the messy, rich and thick data into discernible patterns, concepts, processes or mechanisms. To move from an anecdotal to a *systematic* approach in doing qualitative research, researchers can utilize the advances in computing technologies in the form of CAQDAS and adopt one of the dominant qualitative approach methodologies (see Chaps. 1 and 3). In this chapter, we describe the rationale for using CAQDAS in qualitative research as a strategy to professionalize and legitimize qualitative research. We also compare RQDA — the R package for Qualitative Data Analysis — the CAQDAS we introduce here, with three of the most well-known CAQDAS software (One of the earliest CAQDAS tools was NUDIST (Richards T, & Richards L, Qual Sociol 14(4):307–324, 1991). Other types of CAQDAS software, which are proprietary based (requires license fees), are MAXqda, QDA Miner, Dedoose, and webQDA. Open source (free) CAQDAS software includes Aquad, Coding Analysis Toolkit (CAT), Transana, and WeftQDA, among others) programs, NVivo, ATLAS.ti, and MaxQDA.

Keywords CAQDAS · Qualitative research · RQDA · ATLAS.ti · NVivo · MaxQDA

Y. Chandra, L. Shang, *Qualitative Research Using R: A Systematic Approach*,
https://doi.org/10.1007/978-981-13-3170-1_2

2.1 A Review of CAQDAS-Based Qualitative Research

CAQDAS-based research has been practiced for more than 30 years. Researchers have used CAQDAS to conduct research in various fields including psychology, sociology, management, and political science, to health sciences[1]. Qualitative researchers should be aware that CAQDAS software is different from software that performs automated text mining or automated content analysis tools such as Leximancer, ALCESTE or KH Coder. These tools automatically plot a body (corpus) of textual data analyzed by a software program to produce qualitative outputs, with little to no researcher input. Instead, CAQDAS-based qualitative research requires researcher input in terms of giving meaning and coding a body of textual data. CAQDAS software also differs from the "bag of words" software, such as Linguistic Inquiry and Word Count (LIWC) or DICTION, which converts textual data into numbers for statistical analysis.

CAQDAS software facilitates rigorous, systematic, defendable and auditable qualitative research, but it does not replace researchers' *intelligence, creativity, diligence,* and *attention to detail* as researchers must interpret the *data, mark* and *label the code,* and *display the code* with the assistance of CAQDAS software.

Before CAQDAS software became ubiquitous, qualitative scholars conducted qualitative research *manually*. That is, they used highlighter markers, pens/pencils, multi-colored Post-it notes to label and mark qualitative data (e.g., transcribed interviews, organizational documents, news articles) and sort it into meaningful categories. However, the *manual* approach to doing qualitative research has some disadvantages, which is why researchers shifted to CAQDAS-based approach. For instance, the manual approach limits the ability of the researchers to *share* the coded materials with research collaborators (especially in projects involving international collaborators) and gatekeepers (i.e., journal editors and their reviewers). When the research project is large (containing dozens or hundreds of transcribed interviews), the analytical coding work can become very messy which makes it difficult to *go back and forth* between data and codes when a researcher seeks to make sense of and report the findings. It is also easy to *overlook* at important details in the data. Importantly, it limits the *retrieval* of coded materials.

A key difference between qualitative research using CAQDAS and a manual approach is the *coding* process. A manual approach to the qualitative coding process is illustrated in Fig. 2.1. As shown in Fig. 2.1, researchers using a manual approach using different colors of Post-it notes and highlighters to mark and code different concepts embedded in qualitative data (see the left panel). When combined the

[1]CAQDAS software has been used to facilitate qualitative research in areas such as arts and humanities (Norgaard 2011; Paulus and Lester 2016), business and management (Molecke and Pinkse 2017; Wright and Nyberg 2017), non-profit management and social enterprises (Brown and Guo 2010; Chandra and Shang 2017), social work (Oswald 2017), urban and environmental studies (Lozano and Huisingh 2011; Moore 2009), to public administration and political science (Fisher 2013; Micheli and Neely 2010), as well as health sciences (Colgrove et al. 2010; Håkanson et al. 2010).

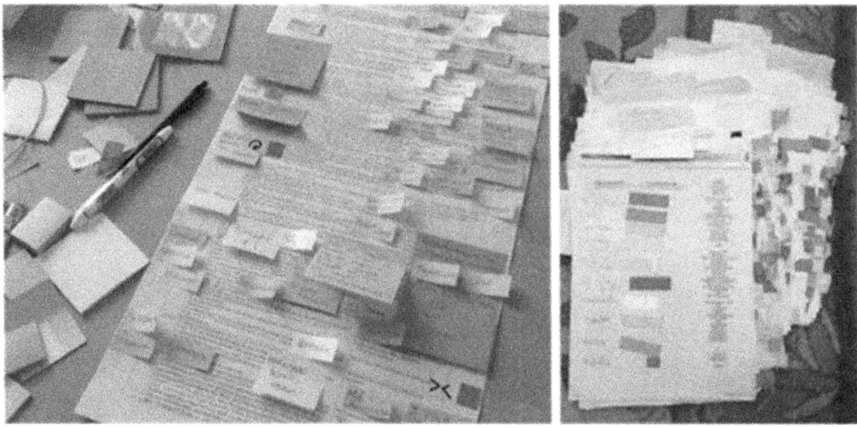

Fig. 2.1 A coding process using the manual approach. (Contributed by Dr. Rikke Haller Baggesen from Roskilde University, Denmark)

coded materials may look like the one on the right panel. When the project is relatively large (a large amount of data collected or a longitudinal research design), a researcher may need manila folders and plastic containers to separate them into meaningful categories.

As shown in Fig. 2.2 below, we demonstrated how a paragraph of the strategies and activities of a social innovator17 can be coded into multiple concepts (as shown on the left screen), using the RQDA software. When each of the coded concepts is clicked, it will show what textual data is classified under each concept. Coders can write memos for each file coded (e.g., a transcribed interview) and for each code (using Attributes function, see Chap. 9).

We claim that CAQDAS enhances and strengthens the qualitative research process along the lines advocated by qualitative methodologists, from Miles and Huberman (1994), Strauss and Corbin (1990) to Gioia et al. (2013).

Below are some important reasons for using CAQDAS software in qualitative research:

- It enhances the *transparency* and therefore the *trustworthiness* of a qualitative research through visible and share-able coded outputs (Woods et al. 2015)
- It increases the efficiency of the *coding process* — making and labeling codes, amending/re-labeling/deleting codes — and in *comparing codes* (Bazeley 2007; Saillard 2011)
- It enables *finer-grained coding*, or *line-by-line coding* (Creswell 1998)
- It supports deeper levels of analysis than is possible with manual coding (Paulus and Lester 2016)
- It facilitates coding to *multiple categories* and deeper exploration of the data (Wickham and Woods 2005)
- It enables researchers to check *coding consistency* (Weber 1985)

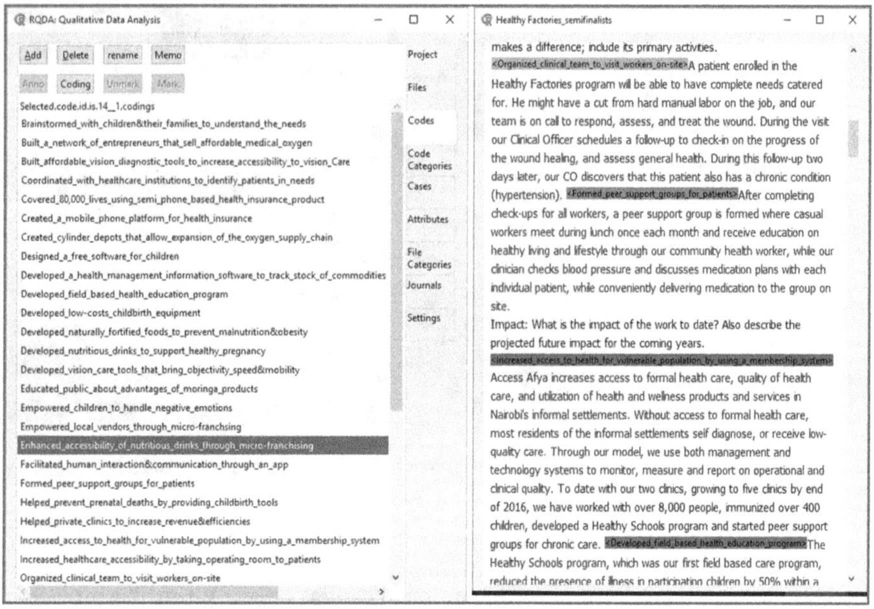

Fig. 2.2 A coding process using RQDA

- It provides '*memo-ing*' functions to record emergent theoretical ideas as well as coding procedures (Babbie 2002)
- It enhances the efficiency of the *coded data management* of relatively large data sets (Morse and Richards 2002)
- It eases the *search* and *retrieval* of coded data (Sinkovics and Alfoldi 2012).
- It eases the coded data *plotting* (visualization)
- It provides a more professional image for qualitative research (Rodik and Primorac 2015)
- It has gained legitimacy with increasing numbers of CAQDAS-based qualitative papers *published in top-tier journals*
- It enables *mixed methods research* for some CAQDAS software, including RQDA

Nevertheless, qualitative researchers should also be aware of the potential pitfalls of CAQDAS software for qualitative research:

- It creates a lack of awareness (called *reflexive blindness*) of how CAQDAS software influences the way researchers conduct research (Woods et al. 2016)
- It makes data analysis a *rigid and inflexible* process (MacMillan 2005)
- It contextually *distances* a researcher from his/her the data (Sotiriadou et al. 2014)
- It causes a researcher becoming too close to the data, called a *coding trap* (Gilbert 2002)
- It provides *technical obstacles* in learning and operating CAQDAS software

- It is less "natural" compared to using paper and pencil (Rodik and Primorac 2015).
- It assumes a *positivistic* approach, which sees the natural world as composed of objects that humans can study, understand and manipulate (Roberts and Wilson 2002) and therefore contradicts with the interpretivist roots of qualitative research

Some researchers resist CAQDAS due to budgetary constraints. Major proprietary CAQDAS software programs are often expensive for individuals, and in this era of university and research institute budget cutting, resources to purchase software are often limited. Proprietary CAQDAS software can easily cost over US$700 for a single user with an educational license. This is why RQDA, the software used in this book, will provide a solution to such problems because it is open source (free!) and is regularly maintained by its developer.

In a nutshell, CAQDAS is like a knife: its value depends on the user's goals, how it is used and, and the beliefs of the community of users about what it is and how it should be used (or not).

2.2 Major CAQDAS Software and the Emergence of RQDA

Selecting a CAQDAS software program is like buying a smartphone: it depends on one's needs and goals. One often needs to trade off one attribute for another such as choosing a CAQDAS that can support different types of data files but has a jargon-filled manual and complex software terms. Or some CAQDAS programs offer around-the-clock technical support, but at considerable cost. To assist readers to make an intelligent choice, we offer a comparison among four popular CAQDAS software programs — three proprietary programs, Atlas.ti, NVivo, and MaxQDA and RQDA, the free open-source CAQDAS software of our interest. The comparison focuses on three aspects: *technical specifications*, *computing capabilities* and *methodological orientation* (see Table 2.1).

As shown in Table 2.1, RQDA is the most recently released CAQDAS software in the market. The rest were developed in the late 1980s but regularly updated. RQDA is one among the few CAQDAS that is *free of charge*. It may remain free because it is maintained by a single developer who does not seek profits from the software. The other three programs are developed and maintained by companies that charge license and training fees.

RQDA is developed as a *package*, as part of the highly popular open-source R computing platform. As open source software, RQDA does not offer any technical support (e.g., no online support or phone hotline center) from its developer.

To date, RQDA supports only text (.txt[2]) files and does not support audio, images, videos, and social media files as commercial CAQDAS software do. However, it

[2] For Windows users, a researcher can open a qualitative data document (e.g., transcribed interview) and then click "save as" and save it as ".txt" file format. For Mac OS users, click "save as" and save it as ".txt" plain text format.

Table 2.1 A comparison of CAQDAS tools

	RQDA	ATLAS.ti	NVivo	MaxQDA
Technical specification				
First release	2008	1993 (prototype in 1989)	1999 (preceded by NUD*IST, 1989)	1989
Most recent update	2016 (RQDA version 0.2–8)	2017 (Atlas.ti version 8.1)	2017 (Nvivo version 11.4.1)	2017 (MaxQDA version 18.0)
Developer	Huang Ronggui (a sociologist)	ATLAS.ti Scientific Software Development GmbH	QSR International Pty Ltd	VERBI GmbH
License and fee requirements	Free (it runs on R platform)	Proprietary	Proprietary	Proprietary
Availability and type of technical support	Free (open source), but no technical support available	Paid version and technical support is available.	Paid version and technical support is available.	Paid version and technical support is available.
Operating system supported	Win, OSX, Linux	Win, OSX, iPad, Android	Win, OSX	Win, OSX
Supporting software needed	RQSlite for database; RGtk2 and GTK+ for graphical user interface	None	None	None
File compatibility	Text (.txt) only	Audio, graphic, text, video	Audio, graphic, text, video	Audio, graphic, text, video
Languages supported for data analysis	Allows import data in several foreign languages	Allows import data in several foreign languages	Allows import data in several foreign languages	Allows import data in more than 10 foreign languages
Capabilities				
Programming language skills requirements	Needs some knowledge of R programming language, also have point and click function	No programming language skills needed, it's a point and click program	No programming language skills needed, it's a point and click program	No programming language skills needed, it's a point and click program
Statistical functions and mixed methods ability	Enables users to write R commands for statistical analysis and apply various R packages for statistical analysis "under one platform"	Enables data attributes to be transformed into tabulations or cross-tabulations (matrix) for further statistical analysis in SPSS	Enables data attributes to be transformed into tabulations or cross-tabulations (matrix) for further statistical analysis in SPSS	Allows descriptive statistical analysis (for MaxQDA Analytics Pro only)

(continued)

Table 2.1 (continued)

	RQDA	ATLAS.ti	NVivo	MaxQDA
Capacity for simultaneous, multi users use	No simultaneous use of software by multiple users	Lack of simultaneous use of software by multiple users	Allows simultaneous use of software by multiple users	Limited simultaneous use of software by multiple users
GUI interface	Simple and easy to navigate for novice and experienced users	Somewhat complex to navigate for novice users; too many buttons and menu functions	Somewhat complex to navigate for novice users; too many buttons and menu functions	Somewhat complex to navigate for novice users; too many buttons and menu functions
Primary CAQDAS functions	Coding, Aggregation, Query, Memo, Visualisation	Coding, Aggregation, Query, Memo, Visualisation	Coding, Aggregation, Query, Memo, Visualisation	Coding, Aggregation, Query, Visualisation, Statistical Tools
Capabilities in code aggregation	Up to two levels of hierarchical structure of coding (simplicity is king)	Limited to no function for hierarchical structure of coding	Good function for hierarchical structure of coding	Good function for hierarchical structure of coding
Codes output sharing	Can be exported to a HTML file	Can be exported to SPSS and XML	Can be exported RTF, Excel and HTML table	Can be exported Excel, HTML and RTF
Boolean search	Allows for complex Boolean search	Allows for complex Boolean search	Allows for complex Boolean search	Allows for complex Boolean search
Type of outputs produced	Aesthetically looking visual plots of first and second order codes	Code list, code hierarchy, code co-occurrence, code neighbors	Chart coding, word tree, word cloud, mind maps, concept maps, cluster analysis, tree maps, sociograms	Code matrix, code relations, document portraits, document comparison chart, word cloud
Methodological				
Treatment of all materials as one hermeneutic unit	Yes, only .txt files	Yes, all types of files	Yes, all types of files	Yes, all types of files
Large data set recording, storage, indexing, sorting	Yes	Yes	Yes	Yes

(continued)

Table 2.1 (continued)

	RQDA	ATLAS.ti	NVivo	MaxQDA
Cross border collaboration	Yes, via file sharing	Yes, real time team collaboration	Yes, real time team collaboration	No real time team collaboration
Coding flexibility	Yes	Yes	Yes	Yes
Codes comparison	Yes	Yes	Yes	Yes
Exhaustive coding schemes	Yes	Yes	Yes	Yes
Researcher reflexivity needed	Yes	Yes	Yes	Yes
Closeness to data	Yes	Yes	Yes	Yes
Data attributes addition	Yes	Yes	Yes	Yes
Transparency and credibility of data analysis	Yes	Yes	Yes	Yes
Efficient, accurate retrieval	Yes	Yes	Yes	Yes
Visualization of codes	Yes	Yes	Yes	Yes

supports text data in more than 10 non-English languages, including English, Mandarin, Japanese, Korean, Arabic, German, French, Spanish, Indonesian and other major languages, and it supports PDF documents that are highlighted in color.

RQDA is designed for *coding simplicity*, and thus deliberately avoids using complex tree-like code categorizations. It offers a *two-level* code aggregation, that is, a first-order code and second-order themes to support theory-building research. The rationale for the simple and one or two-level data aggregation function in RQDA is to enable researchers to drive the theory-building process (Denzin and Lincoln 2005).

A first-order concept is a representation or summary of a text (e.g., a phrase, a sentence, a paragraph, or multiple paragraphs of an in-depth interview) coded into a short label. For instance a sentence from an interview transcript "we employed ex-terrorists to do all kinds of service work, from serving to waiting on customers. Why is this important? Because it helps them be more humble, as they think they are the most righteous" can be interpreted and coded it as: "**serving_waiting_customers_as_humbling**" as a first-order concept. When a researcher has coded a grouping of first-order codes, he/she will then aggregate those with similar meaning into a higher-order theme called a *second-order theme*. For instance, first-level codes such as “interacting with people of different religions, races and ethnicities”, “interacting with state officials”, and “interacting for reconciliation with families of terror victims” can be aggregated to “**broadening_social_circles**” as a second-order theme.

The advantage of RQDA is that it runs on the R computing platform. RQDA can be extended to include statistical analysis of coded results (a more *positivistic* approach to doing qualitative research; see Chap. 1) and thus enables *mixed method research* in a single platform. Therefore, RQDA can be easily adopted by experienced R users. For novice users, RQDA opens a new window of opportunity to enter the world of R as a powerful computational tool. Step-by-step instructions for new users are presented in Chaps. 5, 6, 7, 8, 9, 10, and 11.

Although it takes some basic programming skills to start and launch RQDA, the programming skill required is very basic, and users can store the syntax in a non-programmable file (e.g., notepad or .txt file) and easily copy-and-paste it to R console to avoid errors. For coding work, like the ATLAS.ti, NVivo and MaxQDA programs, RQDA is a fully GUI (graphical user interface)-based program. The actual uses of RQDA — from new project creation, importing transcribed interviews, and line-by-line coding, to code labeling, aggregating, and plotting — is similar to the other three software programs through “point and click” in a GUI. Importantly, the GUI for RQDA is much simpler than for the other software. This makes it less intimidating for novice users. Over the years, with better technical functionalities and capabilities in ATLAS.ti, NVivo and MaxQDA, their GUI has also grown in complexity which can “turn off” novice users. Less is more.

The visualization of coding output in RQDA is also relatively simple and the output looks aesthetic and professional. Unlike the other software that has many output visualization capabilities, RQDA focuses on a two-level code visualization, thus making it easy to read and to “identify the story”.

The transformation of coded outputs into an HTML file (or more than one HTML files) is another advantage of RQDA. It allows the researchers to share the coded data with collaborators across geographical borders. It also enables easy keyword searches on codes or key words of interest in the HTML file.

From a methodological perspective, RQDA is very similar to ATLAS.ti, NVivo and MaxQDA in that all textual data (e.g., interview transcripts, news articles, organizational documents) imported into the software are treated as a single entity for analysis or *one hermeneutic unit*. RQDA, like the other three comparison programs, can record, store, sort and retrieve large coded datasets, and facilitates flexible coding (amending and deleting code labels, etc.). It also creates audit-able workflow so that other researchers or reviewers can check how data were analyzed and the consistency in the coding process. Furthermore, it can record data attributes (e.g., source and year of data, demographic variables of interviewees) and comparison of cases based on the data attributes to provide additional insights into the data.

2.3 Conclusion

As Yin (1994, p. 16) said, "paradoxically, the 'softer' a research strategy, the harder it is to do." As a soft science, qualitative research has all the benefits of incorporating CAQDAS as a strategy to address many of the questions that the researcher, along with readers and reviewers, often raise about the quality of a qualitative research. CAQDAS software offers a golden opportunity to enhance the legitimacy and power of qualitative research. In this chapter, we have explained the merits of RQDA and identified how it is similar to and different from mainstream CAQDAS software. Just as technologies advance medical practice, we believe that they will also advance (and professionalize) qualitative research practices.

References

Babbie, E. (2002). *The basics of social research* (2nd ed.). Belmont: Thomson Learning.

Bazeley, P. (2007). *Qualitative data analysis with NVivo*. London: Sage.

Brown, W. A., & Guo, C. (2010). Exploring the key roles for nonprofit boards. *Nonprofit and Voluntary Sector Quarterly, 39*(3), 536–546.

Chandra, Y., & Shang, L. (2017). Unpacking the biographical antecedents of the emergence of social enterprises: A narrative perspective. *Voluntas: International Journal of Voluntary and Nonprofit Organizations, 28*(6), 2498–2529.

Colgrove, J., Abiola, S., & Mello, M. M. (2010). HPV vaccination mandates—Lawmaking amid political and scientific controversy. *New England Journal of Medicine, 363*(8), 785–791.

Creswell, J. W. (1998). *Qualitative inquiry and research design: Choosing among five traditions*. London: Sage.

Denzin, N. K., & Lincoln, Y. S. (2005). Introduction: The discipline and practice of qualitative research. In N. K. Denzin & Y. S. Lincoln (Eds.), *The Sage handbook of qualitative research* (3rd ed., pp. 1–32). Thousand Oaks: Sage.

Fisher, D. R. (2013). Understanding the relationship between subnational and national climate change politics in the United States: Toward a theory of boomerang federalism. *Environment and Planning C: Government and Policy, 31*(5), 769–784.

Gilbert, L. S. (2002). Going the distance: 'Closeness' in qualitative data analysis software. *International Journal of Social Research Methodology, 5*(3), 215–228.

Gioia, D. A., Corley, K. G., & Hamilton, A. L. (2013). Seeking qualitative rigor in inductive research: Notes on the Gioia methodology. *Organizational Research Methods, 16*(1), 15–31.

Håkanson, C., Sahlberg-Blom, E., & Ternestedt, B. M. (2010). Being in the patient position: Experiences of health care among people with irritable bowel syndrome. *Qualitative Health Research, 20*(8), 1116–1127.

Lozano, R., & Huisingh, D. (2011). Inter-linking issues and dimensions in sustainability reporting. *Journal of Cleaner Production, 19*(2), 99–107.

MacMillan, K. (2005). More than just coding? Evaluating CAQDAS in a discourse analysis of news texts. *Forum: Qualitative Social Research, 6*(3), Art. 25. Retrieved from http://www.qualitative-research.net/fqs/fqs-eng.htm

Micheli, P., & Neely, A. (2010). Performance measurement in the public sector in England: Searching for the golden thread. *Public Administration Review, 70*(4), 591–600.

Miles, M. B., & Huberman, A. M. (1994). *Qualitative data analysis: An expanded source book* (2nd ed.). Thousand Oaks: Sage.

Molecke, G., & Pinkse, J. (2017). Accountability for social impact: A bricolage perspective on impact measurement in social enterprises. *Journal of Business Venturing, 32*(5), 550–568.

Moore, K. S. (2009). Gentrification in Black face?: The return of the Black middle class to urban neighborhoods. *Urban Geography, 30*(2), 118–142.

Morse, J. M., & Richards, L. (2002). *Read me first for a user's guide to qualitative methods*. Thousand Oaks: Sage.

Norgaard, M. (2011). Descriptions of improvisational thinking by artist-level jazz musicians. *Journal of Research in Music Education, 59*(2), 109–127.

Oswald, A. G. (2017). Improving outcomes with qualitative data analysis software: A reflective journey. *Qualitative Social Work*, 1473325017744860. https://doi.org/10.1177/1473325017744860

Paulus, T. M., & Lester, J. N. (2016). ATLAS.ti for conversation and discourse analysis studies. *International Journal of Social Research Methodology, 19*(4), 405–428.

Richards, T., & Richards, L. (1991). The NUDIST qualitative data analysis system. *Qualitative Sociology, 14*(4), 307–324.

Roberts, K. A., & Wilson, R. W. (2002). ICT and the research process: Issues around the compatibility of technology with qualitative data analysis. *Forum Qualitative Sozialforschung/ Forum: Qualitative Social Research, 3*(2), Art. 23. http://nbn-resolving.de/urn:nbn:de:0114-fqs0202234. Accessed 8 Jan 2013.

Rodik, P., & Primorac, J. (2015, January). To use or not to use: Computer-assisted qualitative data analysis software usage among early-career sociologists in Croatia. In *Forum: Qualitative social research* (Vol. 16, No. 1). Berlin: Freie Universität

Saillard, E. K. (2011). Systematic versus interpretive analysis with two CAQDAS packages: NVivo and MAXQDA. *Forum Qualitative Sozialforschung/Forum: Qualitative Social Research, 12*(1). http://www.qualitative-research.net/index.php/fqs/article/view/1518

Silverman, D. (1993). *Interpreting qualitative data: Methods for analysing talk, text, and interaction*. London: Sage.

Sinkovics, R. R., & Alfoldi, E. A. (2012). Progressive focusing and trustworthiness in qualitative research. *Management International Review, 52*(6), 817–845.

Sotiriadou, P., Brouwers, J., & Le, T. A. (2014). Choosing a qualitative data analysis tool: A comparison of NVivo and Leximancer. *Annals of Leisure Research, 17*(2), 218–234.

Strauss, A., & Corbin, J. M. (1990). *Basics of qualitative research: Grounded theory procedures and techniques*. Newbury Park: CA, Sage.

Weber, P. R. (1985). *Basic content analysis*. Beverly Hills: Sage.

Weick, K. E. (1989). Theory construction as disciplined imagination. *Academy of Management Review, 14*(4), 516–531.

Wickham, M., & Woods, M. (2005). Reflecting on the strategic use of CAQDAS to manage and report on the qualitative research process. *The Qualitative Report, 10*(4), 687–702.

Woods, M., Paulus, T., Atkins, D. P., & Macklin, R. (2015). Advancing qualitative research using qualitative data analysis software (QDAS)? Reviewing potential versus practice in published studies using ATLAS.ti and NVivo, 1994–2013. *Social Science Computer Review*, 0894439315596311.

Woods, M., Macklin, R., & Lewis, G. K. (2016). Researcher reflexivity: Exploring the impacts of CAQDAS use. *International Journal of Social Research Methodology, 19*(4), 385–403.

Wright, C., & Nyberg, D. (2017). An inconvenient truth: How organizations translate climate change into business as usual. *Academy of Management Journal, 60*(5), 1633–1661.

Yin, R. (1994). *Case study research design and method*. Beverly Hills: Sage.

Chapter 3
How to Conduct Caqdas-Based Qualitative Research

Abstract In this chapter, we operationalize the interpretivist-constructivist Gioia approach to qualitative research by integrating it with RQDA and demonstrating how to do so.

Keywords Gioia qualitative research model · RQDA · CAQDAS-based qualitative research · Data collection · Data analysis

3.1 Step 1: Articulating a Research Question

As a best practice approach to research, one should start a qualitative study by asking a clear, useful and meaningful *research question*. This is necessary not only because gatekeepers and readers often look for it but also because articulating a research question (1) helps the researcher to remain *focused* to answer a question thoroughly (especially given the 'messy' and lengthy nature of qualitative research), and (2) makes it *easier* for reviewers and journal editors to readily understand a qualitative paper and save time in checking if a manuscript has answered the question(s) asked (see Table 3.1).

Qualitative researchers often wrestle with these questions when formulating their research question: (1) What kind of questions are appropriate for qualitative research? (2) What is the acceptable number of research questions I should have in a manuscript? (3) How should I frame a research question?

First, qualitative research is best designed to answer the *why* and *how*[1] question(s). 'Why' and 'how' questions typically require a 'long' versus 'short' answer, which meshes with the open-ended nature of qualitative data. The questions often require

[1] For example, in a study to understand how citizens of Sicily, Italy resisted paying protection money to mafia organizations, Vaccaro and Palazzo (2015) asked a 'how' question: *How can change initiatives succeed in social contexts that are dominated by organized crime*? In another study, Gioia et al. (2010) explored the processes involved in forming an organizational identity using two 'how' questions: How do members of *a nascent organization develop a collective understanding of "Who are we as an organization?" and How does a newly created organization develop a sense of itself as a social actor in its field or industry?"*

Y. Chandra, L. Shang, *Qualitative Research Using R: A Systematic Approach*,
https://doi.org/10.1007/978-981-13-3170-1_3

Table 3.1 The Gioia qualitative research model using RQDA

Step	Key guidelines	Example from Chandra (2017)
(1) Articulate a research question	Articulate a clear research question at the start	"How can social entrepreneurship emancipate individuals from the constraints of ideology and their own past behavior?"
	Pose '*why and how*' questions about change, process, or how things operate and unfold	
	Not more than two questions per qualitative manuscript	
	New and meaningful question(s) desirable	
(2) Conduct initial literature review	Consult the literature at the outset for background information and gaps but do not let this constrain or guide your investigation.	Read a news article written by a reformed ex-terrorist who helped other terrorists leave the world of terrorism and become an entrepreneur. This was a "wow" moment leading to some intriguing questions: "Can ex-terrorists really be rehabilitated? Why and how is that possible?"
	Refine the research question after initial literature review if necessary.	
	Explore literature that offers a *counterintuitive* perspective to the phenomenon of interest.	Cognizant of terrorism theories and critical social theories (e.g., emancipation, oppression, ideology) and psychological theories (e.g., identity, self-esteem, empowerment).
	Import or borrow literature from other disciplines to extend/enrich existing theory/knowledge.	
	Focus on a *new context* thus explore new literature to extend and enrich existing theory/knowledge.	
(3) Perform sample selection	Use *purposive sampling* (to find samples that can offer the best contributions to the theory/practice/policy), or move from purposive sampling to *theoretical sampling* (by doing preliminary data analysis and then decide what data to collect next, until a theory emerges).	Findings that ex-terrorists who became positive change agents could help other ex-terrorists find an alternative lifestyle was a *revelatory case* (previously inaccessible to social scientists); thus worthy of investigation. The choice to study these ex-terrorists was intentional, to illuminate new phenomenon and processes (purposive sampling). Due to a lack of other similar cases for comparative processes, the researcher decided to focus on the change agent ex-terrorists.
	Focus on achieving *analytical generalization*.	
	Pragmatism: multiple-case design is desirable, but single-case design is useful for various rationales (i.e., *novel case, revelatory case, extreme case, deviant case*, or *typical case* for comparative purposes).	
	For multiple-case design: employ *literal replication* to help predict similar results; *theoretical replication* to produce contrary results for predictable reasons.	
	Ensure the case site is legitimate, trustworthy, and offering new insights.	

(4) Craft an interview protocol	Craft an *initial* interview protocol; *adjust* the interview questions as the data collection progresses to capture new, important themes or ideas.	Interview protocol contained questions such as "Why and how did you join terrorism?", "What made you leave the world of terrorism?", "How did you start helping other terrorists to live a normal life?"
	Semi-structured interview protocol is preferred.	
(5) Conduct data collection process	Conduct multiple interviews and use multiple informants to achieve the best results.	Interviews took place between 2013 and 2016 with 7 individuals including ex-terrorists, staff members of a social enterprise, terrorism experts, in which 2 individuals were with ex-terrorists (total = 10 interviews).
	Use *data triangulation* if possible.	
	Focus on collecting *rich* and *thick* data.	Secondary data collected include 57 online news articles, YouTube and TedTalk videos, and a documentary made by one of the ex-terrorists.
	Data collection and analysis timing overlap.	
	Be inclusive by treating "all as data" (code the primary interview transcripts, news articles, observations, archival data etc.).	
(6) Perform data analysis	**Data analytic preparation**	
	1. Install R, RSQLite, GTK+, RQDA on a computer.	
	2. Save qualitative data as individual ".txt" files.	
	3. Launch RQDA software.	
	4. Create a new project.	
	5. Import all .txt files into RQDA software.	
	Coding process	A main researcher acted as a data coder; other researcher(s) critiqued the coding results; sought consensus on the coded data.
	1. In RQDA, perform inductive data analysis by coding the .txt data into *first-order concepts*. Read, interpret data and click "Mark" *("Unmark")* on imported data to do (undo) the coding.	
	2. Click "Attributes" and capture key data attributes (e.g., data sources, informant demographics, dependent variables).	Presented the findings in local and international conferences to seek feedback to refine the coding results.

(continued)

Table 3.1 (continued)

Step	Key guidelines	Example from Chandra (2017)
	3. Click "Memo" to write memos for each case; the memos can be on the content or decisions or processes in the coding.	Some codes were relabeled to refine meaning; other codes that were similar in meaning were further merged for parsimonious reason.
	Code aggregation	
	9. Use "Add To" button to aggregate all first-order concepts to *second-order themes*. To further aggregate second-order themes to aggregate dimensions, copy and paste all (first-order and second-order) codes to Excel spreadsheet and combine those with similar meaning to create higher level codes accordingly.	
	Code plotting and sharing	Shared the coded HTML outputs with outside researchers via email and or printed copies to seek feedback.
	10. Select all second-order themes using cursor, click "right mouse" and press "Plot Selected Code Categories."	
	Showcase the findings	Developed an analytical coding process/data structure to summarize the findings (see Fig. 3.1).
	11. Re-arrange all coded data into a "data structure" format (using a table to organize them) and or using three layers of analytical coding process.	
(7) Articulate a theory	Articulate the dynamic relationships among second order and or aggregate dimensions into a process model by going back and forth (as many times as necessary) between findings and the literature to develop your own theory.	Achieve 'reach closure' by developing a process model (see Fig. 3.2)

an explanation of a process, event or mechanism. Second, it is common for researchers to design qualitative research that contains multiple research questions, especially in a longitudinal or complex project. However, when writing a manuscript for journal publication, the gatekeepers—editors and reviewers—almost always want to see *one or two research questions*[2]. If the researcher has two questions, usually they are deeply interrelated. Third, researchers also need to frame a research question so that it is appropriately *new and meaningful* yet not too narrow or too broad to restrict or blur a researcher's contribution to a field. For instance, Vaccaro and Palazzo (2015) framed a question that contained a novel context (e.g., mafia-controlled society in Sicily, Italy) from the perspective and actions of anti-mafia activists, and tied the study to the literature on institutional change, and specifically, the role of values. The study added value to the field by theorizing that the strategic use of 'values work' is an instrument to institutionalize change in a society. In another study, Chandra (2017) framed a research question to extend the theory of emancipation by investigating the behavior and liberation of former political terrorists as individuals driven by ideology. This work advanced the research on social entrepreneurship by offering a new process model and perspective of social entrepreneurship as emancipatory work.

3.2 Step 2: Conducting a Literature Review

The process of finding, sharpening and framing a research question(s) often involves an *initial* literature review of a topic or field of research interest. The process is cyclical (back and forth) rather than linear to ensure that a research question(s) is *appropriate*[3] for qualitative study and that the question asked can make an original and strong *contribution* to a field. As a consequence of moving back and forth between literature and research questions, a researcher may *change* an original research question, or *refine* a research question, or simply *reframe* its perspective. When conducting qualitative research, a researcher could observe an *unexpected* yet very interesting phenomenon during the research and then decide to re-frame the question in the middle of a research project.

Writing the initial literature review in a research manuscript often sounds a lot easier than the actual process. Various strategies have been successfully used by different scholars. Some researchers adopt a *counterintuitive* approach whereby they examine the literature and findings of an area of interest and then ask a research question that counters what is conventionally known it. For example, when there is scholarly consensus about the positive aspect(s) about a phenomenon or construct (e.g., social entrepreneurship and its positive social impact), a researcher may want

[2] Some researchers ask three to five research questions in a qualitative manuscript; this can distract from the paper's focus and positioning and make the paper overly long and lead to diffuse answers.

[3] One way to quickly assess whether a topic deserves another qualitative study is to do a simple Web of Science, Scopus or Google Scholar check to see how many articles have been published on a particular topic area and to read the abstracts of the retrieved articles.

to find literature that contains insights or initial evidence that shows that social entrepreneurship can negatively impact a society. This counter intuitive research approach can be fruitful as it helps researchers develop the *boundary condition* (the condition or circumstances where a theory holds) of a theory or concept.

Another approach is to *borrow* literature from other field(s) to develop a new understanding and explanation in a different field (e.g., bringing in critical social theories to the social entrepreneurship field). Yet another approach is to pick a *new context,* thus new literature to enrich the conversation in a research field (e.g., mafia studies, a topic commonly published in crime studies, to the field of organizational studies and management). The latter two approaches—counterintuitive research and borrowing from other fields—in essence, constitute theoretical innovation.

3.3 Step 3: Selecting a Sample

Should I sample one case or multiple cases? One camp (the qualitative positivist) strictly argues for multiple cases as the best approach to sampling as it facilitates *theoretical replication* as the cases are designed to cover different theoretical conditions that predict contrasting results but for predictable reasons. The other camp argues that a single-case study is legitimate and justified as a means to pursue *analytical generalization* (Yin 2003) where insights from case data lend support to a "theory" (or generalization to "theory"). Clearly, qualitative research does not seek to achieve *statistical generalization*, or generalization to the population (Yin 2003), but to discover new insights or develop new concepts or processes, or *analytical generalization*. Regardless of the approach used, researchers use *purposive sampling* or creatively start with purpose sampling and move to *theoretical sampling* whenever necessary.

We advocate for a pragmatic approach and argue that both multiple-case and single-case designs are justified and useful. The sampling decision *depends* on the accessibility of the cases of interest. Assuming that resources and time are not a constraint, the most obvious constraint is the *availability* of the case (e.g., an individual, organization, group, or country) to be included in qualitative research. One other strategy is to ensure the case site is *legitimate, trustworthy, offering new insights,* and *worth publishing*.

So, when is a single-case design the best or ideal approach? When a researcher *discovers a new phenomenon*, a single case is justified. For instance, think of the first case of a non-Moslem ethnic minority man who became the governor of Jakarta, Indonesia (home to the largest Moslems population worldwide) in 2014; this deserves a single case study as it was the very first case in the Indonesian politics (Hatherell and Welsh 2017). A researcher could ask this question: "What drives the electoral success of ethnic minority politicians in a new democracy?" Another single case design is the emergence of a new type of political communication (e.g., President Donald Trump's controversial verbal acts) and how this form of communication a U.S. leader affects trading relations with its strategic rivals.

Another justified approach to single case sampling is when a phenomenon is *revelatory*—when it offers new access to previously inaccessible scientific investigation. For instance, choosing to sample a social enterprise (e.g., NGO) that targets and empowers ex-terrorists in light of the fact that most researchers do not have access to such organizations (e.g., most ex-terrorists are not willing to be studied; Chandra 2017).

Another reason for employing a single-case approach is when an event is *extreme* or rare or unique; that is, it could offer documentation and analysis of a rare event. For example, the Cuban missile crisis (Allison 1969), one of the most dangerous moments in modern history as part of the US-Soviet Union rivalry, is an extreme case for study. Fukushima power plant accident, genocides by the Nazis, and Western invasion of the Americas are other good examples of extreme cases.

Some researchers select a *deviant* phenomenon, which describes a very unusual phenomenon or event with the purpose of refining or replacing existing theory or hypothesis (usually used to generate new propositions). For instance, McDougall et al. (1994) sampled 24 cases of "born global" firms (i.e., a business organization that, from inception, seeks to derive significant competitive advantage from the use of resources and the sale of outputs in multiple countries) to *challenge* existing internationalization theories that show that firms need to grow and mature in domestic markets before they can become international. This work subsequently led to the *birth* of a new field called *international entrepreneurship*.

Last but not least, the single-case design is justified when a researcher is focusing on *typical* phenomena. Such studies describe what is typical, normal or average so as to provide a deeper understanding of the phenomenon. Many qualitative researchers use this approach, for example they use a single case to highlight already known phenomenon such as 'the pull and push factors in exporting' (the 50-year old research on exporting has revealed much about the push and pull factors that lead a firm to export goods and services), although this design has not identified any major theoretical breakthroughs.

For multiple-case designs, researchers typically employ the *literal replication* approach by selecting a few cases (usually three to four) that predict similar results (e.g., if the first case shows that merger increases firm performance, then a researcher will pick two to three more cases in which firm performance increased after the merger). Literal replication is most often used in tandem with *theoretical replication* (analyzing another three or more cases) as a strategy to produce contrary results for predictable reasons. The aim is to achieve rigorous findings. For instance, to understand why and how certain entrepreneurs *search for*, instead of *discovering* international market opportunities and whether and if so, how prior knowledge of international markets influences the search or discovery behavior, Chandra et al. (2009) sampled and analyzed cases that demonstrated 'search' behavior (literal replication using three cases) as well as cases that demonstrated 'discovery' behavior (theoretical replication using five cases). These approaches enabled the authors to conclude that the approach to finding international market opportunities may be influenced primarily by the extent and type of the entrepreneurs' prior knowledge, among other factors.

3.4 Step 4: Crafting an Interview Protocol

Most qualitative research involves some form of qualitative interview(s)[4]. An interview protocol is usually needed to enhance the quality of the research and to ensure that the questions to be investigated are geared to answering the research question(s). Some researchers adopt a *structured* interview protocol, while others adopt a *semi-structured* protocol. The semi-structured interview protocol is more commonly used by scholars, due to its flexibility, and it is primarily suitable to discover, explore, or develop new or alternative explanations and perspectives of social reality. As researchers enter the field and conduct interviews, they often reframe or refine the questions in the interview protocol to focus directly on their question(s) and thus sharpen their findings.

A researcher should be inquisitive, vigilant and reflexive to assess whether an interview protocol works so as to help answer the research question(s). It is easy to drift away from the core topic, especially if an informant is 'very chatty' and includes many irrelevant responses to the questions asked, or if he/she is not very articulate or reflective.

3.5 Step 5: Collecting Data

We advise researchers to conduct *multiple interviews with each informant* and interview *multiple informants for each case* whenever possible. These can add richness to the data. How many interviews or informants are needed to craft a successful qualitative study? The answer is, it varies, depending on the context and research question(s). For example, Vaccaro and Palazzo (2015) conducted 84 interviews with 35 activists, 9 with entrepreneurs collaborating with [anti-mafia activists], 7 with entrepreneurs who refused to collaborate with the mafia, 8 with consumers, 11 with students, 7 with teachers and school principals, 5 with Mafia criminologists and police officials, and 2 with politicians. In contrast, Chandra's (2017) conducted 10 in-depth interviews involving seven individuals (i.e., five interviews with two ex-terrorists, one interview with each of three staff members of a small business enterprise aimed to help ex-terrorists establish small businesses, and one interview each with two terrorism experts).

Researchers should also embrace *data triangulation* in the data collection process. That is, although in-depth interviews are the primary data used in qualitative research, collecting additional data through documents, observations, news articles, websites, social media postings, or visual information such as photos and videos enhances the data. Data triangulation not only enhances the *richness and thickness*

[4] Some scholars employ the *quantitative* interview style. They interview informants using a structured questionnaire and give ticks to the questionnaire. This type of interview is usually not associated with qualitative research. Such interview is like conducting a survey, in which the data are 'thin' and 'abstract' rather than thick and concrete.

of qualitative research but also increases the *reliability* of the data. For instance, news articles can be used to cross-check the facts related to what an interviewee says about a phenomenon or event of interest. Photos or YouTube videos can add more information about the background of an interviewee, or the perspective of an issue by an interviewee. Social media postings (e.g., Facebook, LinkedIn, Instagram) may also offer additional clues about a topic of investigation, or help the researcher connect with suitable informants. Some interviewees may be very cooperative and willing to share personal or business documents (e.g., personal diaries, business plans, financial statements, organizational charts, resumes, list of contacts and associates), and these help boost the quality and quantity of the data collected. Some researchers use Skype (or other video technology) to conduct interviews and this approach has gradually gained acceptance in major scholarly journals. But whenever possible, if distance or resources pose no limitations, a face-to-face interview is much better as it allows the researcher to observe the phenomenon *in-situ* (e.g., to understand stigma, one can interview street cleaners at work). So the interviewer can more easily interact with the respondent and not be limited to the sometimes spotty or dropped "face time" conversations.

The data collection and analysis usually overlap in time so a researcher may start recording some key observations about the phenomenon of interest in a safe place (e.g., a notebook or a CAQDAS software such as RQDA, or a smartphone notepad).

3.6 Step 6: Analyzing the Data

Data analysis can be done manually (e.g., using paper, pencil, highlighter, and post-it-notes) or via the help of a CAQDAS program (see Chap. 2). The first step is to perform several software installations to allow RQDA to operate (for *software installations*, see Chap. 5). The second next step is to transform all qualitative data (e.g., transcribed interviews, news articles, organizational documents, social media postings) into a ".txt" files (i.e., the *data*) to allow RQDA to read them (see Chap. 6 for more details on *data preparation*). After RQDA is launched, the researcher can create a new project, name and save it in a directory (see Chap. 7 on how to *start and launch* RQDA). After this step, the researcher will be ready to import all .txt files into RQDA for coding.

The most important part of data analysis in CAQDAS-based qualitative research is the *coding* process. Coding here means creating a *label* that best summarizes a sentence, a phrase, a paragraph, or a body of text so that it can be easily identified and recorded. While coding may sound like a subjective process, when it is performed after a researcher understands key concepts in the literature and involves one or more collaborator and/or 'devil's advocates' (someone who challenges the coded results), the process can generate defendable, plausible, and transparent findings.

In a nutshell, a researcher begins this process by coding a body of qualitative data *as close as possible* to the data to create *first-order concepts* (see Chap. 8 for more details on *how to code* in RQDA). Whenever a sentence, phrase, paragraph, or body of text contains multiple meanings, they should be coded as *multiple codes* (e.g., the old man carries his teenage son on his back" can be coded as "the boy is sick", "the old man is strong" or "an ungrateful son"). This process enriches the data analysis. To code first-order concepts, a researcher will read a phrase, sentence or paragraph of imported qualitative data in RQDA, interpret the phrase, sentence or paragraph and click *"Mark"* on the data to code. A researcher may also frequently undo the codes (*"Unmark"*) when the wrong codes are created and to keep refining the code labels. The RQDA program also has functions to capture the attributes of the data (e.g., data sources, informants' demographics, dependent variables) and for the researcher to write memos for each case (e.g., about on the content, or coding decisions or processes) (see Chap. 9 for more details on *data attributes* and *memo-ing*).

A researcher can use code labels creatively to facilitate better analysis. For example, one can differentiate codes that refer to "Activities" versus "Outcomes" using a prefix "A" and "O" in the code labels (e.g., "**A_using_food_for_peace**"; "**O_reduced_religious_conflicts**"). If a researcher makes cross-case comparisons involving multiple-case design, he/she can create a prefix in the code labels to categorize the meaning of a coded label (e.g., use prefix S for nonprofit organizations that have strong performance and W to indicate weak performance as: "**S_social_performance**", "**W_social_performance**"). This will facilitate cross-case comparisons.

After all first-order concepts have been created, the researcher will aggregate them to *second-order themes* (to crystallize the findings by merging first-order codes that have similar meaning) using the *"Add to"* button in RQDA. Subsequently, a researcher can select all second-order themes by clicking on the right mouse button and selecting *"Plot Selected Code Categories;* this will produce networks of codes that link the first-order concepts and their second-order themes (see Chap. 10 for more details on *code aggregation, plotting and exporting*). This process aggregates the findings into a smaller more manageable concepts. To further combine and abstract the second-level codes to higher level codes (or "aggregate theoretical dimensions"), we suggest that researchers copy and paste all RQDA-generated HTML codes to Excel and re-arrange and combine the codes with similar meaning to create higher level codes in a step by step approach in the spreadsheet.

Finally, all of the coded results need to be organized into categories so they can be easily understood by readers. In doing this, a researcher can re-arrange all coded data into a *data structure* format, sometimes called *analytical coding process* (see Fig. 3.1 below). It is very important to use *informants' quotes* (called "power quotes") to highlight a particular viewpoint, incident or event to bring the evidence closer to the readers and gatekeepers. Usually, a researcher will provide informant quotes separate from those shown in the analytical coding process or data structure.

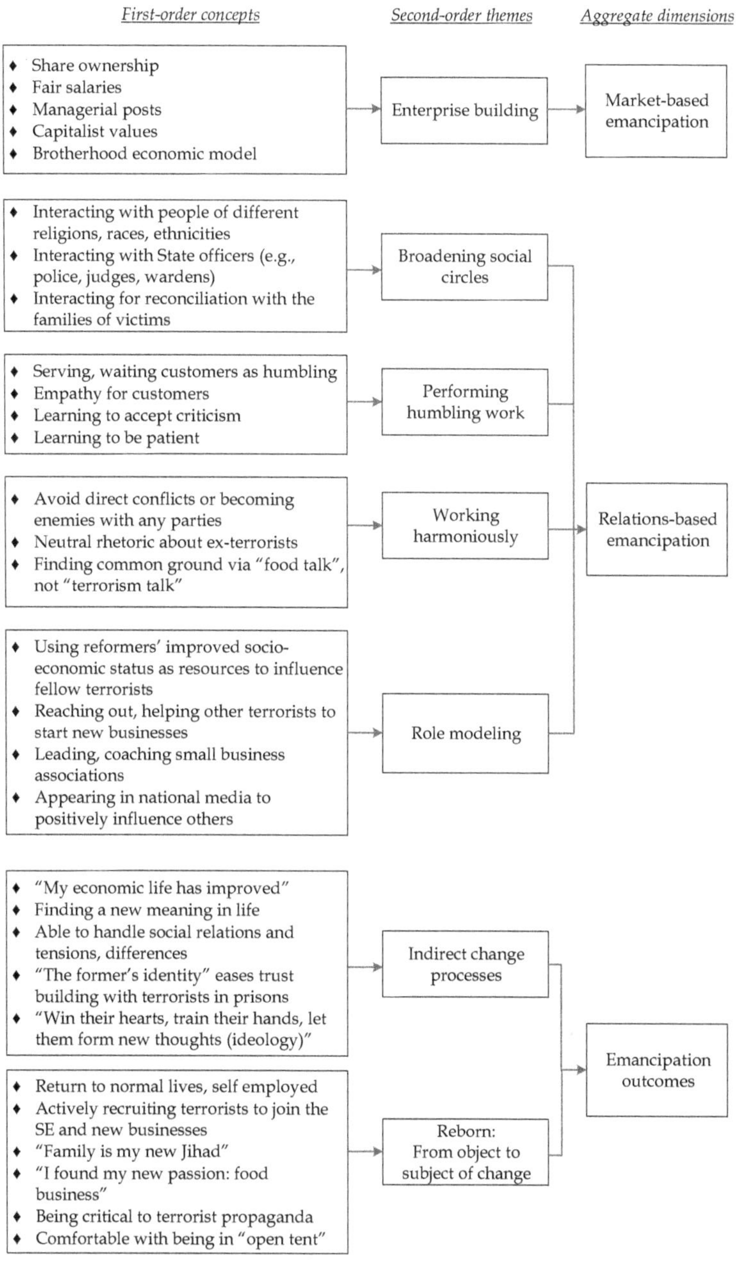

Fig. 3.1 A sample data structure process. (Note: This data structure was created by Chandra (2017) during the journal review process but was not included in the final publication due to word limits)

3.7 Step 7: Summarize Findings with a Theory

The last step of data analysis is to summarize all of the findings to "reach closure". In doing so, a researcher designs a process diagram (or grounded process model) that depicts the relationships among events, activities or constructs identified in the data. This is the stage where the researcher proposes his/her theory about a particular social phenomenon as an answer to the research question(s). Usually, a researcher will write this part by first discussing the theory or process model in general terms, and then explaining them in more details using headings or subheadings and offering the "meat" (the concrete data) to illustrate the workings of the process model.

There are no set rules in crafting a process model or grounded theory, but it is an *iterative* process. Consistency in the use of terminology, striking a balance between simplicity and complexity, highlighting what is novel, unique, or contrary to established beliefs/theory, are critical steps in constructing a powerful "theory" or process model. We encourage researchers to use a professional software (e.g., Microsoft Visio) to draw such a process model. A poorly drawn process model may suggest that a researcher is "lazy" or "lacking due diligence" and this creates a negative impression to journal gatekeepers. Some open source software also works well, such as draw.io. As an example, we depict a process model developed by Chandra (2017) in a study about the emancipation dimensions, processes and outcomes of ex-terrorists (see Fig. 3.2).

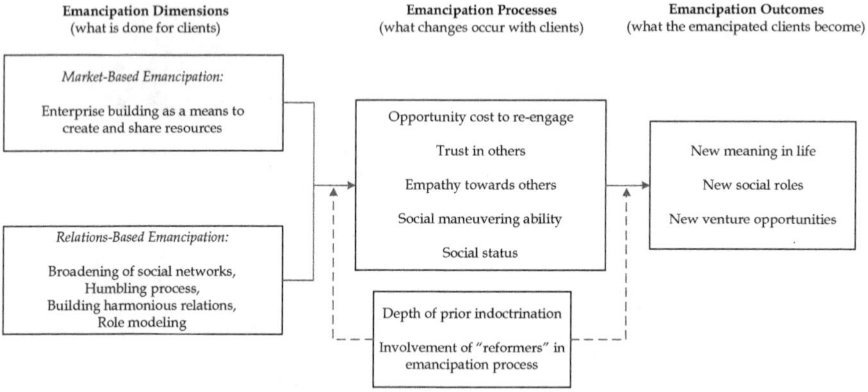

Fig. 3.2 A sample 'Process Model of Emancipation of Ex-Terrorists'. (Source: Chandra 2017)

3.8 Conclusion

In this chapter, we extend the Gioia approach to qualitative research by fusing it with the RQDA software. Although the procedures described in this chapter seem rather linear, they are not because there is room for creativity and improvisation to address research questions and align with or counter respond to research traditions in a particular field. This chapter offers guidance and inspiration for how to conduct qualitative research. There may be other approaches and improvisations that this chapter does not cover.

References

Allison, G. T. (1969). Conceptual models and the Cuban missile crisis. *American Political Science Review, 63*(3), 689–718.

Chandra, Y. (2017). Social entrepreneurship as emancipatory work. *Journal of Business Venturing, 32*(6), 657–673.

Chandra, Y., Styles, C., & Wilkinson, I. (2009). The recognition of first time international entrepreneurial opportunities: Evidence from firms in knowledge-based industries. *International Marketing Review, 26*(1), 30–61.

Gioia, D. A., Price, K. N., Hamilton, A. L., & Thomas, J. B. (2010). Forging an identity: An insider-outsider study of processes involved in the formation of organizational identity. *Administrative Science Quarterly, 55*(1), 1–46.

Hatherell, M., & Welsh, A. (2017). Rebel with a cause: Ahok and Charismatic leadership in Indonesia. *Asian Studies Review, 41*(2), 174–190.

McDougall, P. P., Shane, S., & Oviatt, B. M. (1994). Explaining the formation of international new ventures: The limits of theories from international business research. *Journal of Business Venturing, 9*(6), 469–487.

Vaccaro, A., & Palazzo, G. (2015). Values against violence: Institutional change in societies dominated by organized crime. *Academy of Management Journal, 58*(4), 1075–1101.

Yin, R. (2003). *Case study research: Design and methods*. Thousand Oaks: Sage.

Chapter 4
An Overview of R and RQDA: An Open-Source CAQDAS Platform

Abstract In this chapter, we review the R programming language as a platform for computing in general and we introduce RQDA as an extension of R for qualitative research. We discuss the merits of R and particularly RQDA as an open-source free tool to conduct qualitative research. We provide brief examples of how to use syntax to operate RQDA in R. RQDA's console and its various functions are described for new project creation, data import, and data export, among other functions.

Keywords R · RQDA · R package · R community · R syntax

4.1 An Overview of R

R is a free language and software program for statistical computing and graphics (http://www.r-project.org). As social scientists, we are familiar with statistical packages such as SPSS, SAS, and Stata. R is a powerful alternative to these traditional statistical packages because it enables users to freely distribute, study, change and improve the software under the Free Software Foundation's GNU General Public License, which SPPS, SAS and Stata do not offer. R was created as an implementation of the S programming language by Ross Ihaka and Robert Gentleman at the University of Auckland, New Zealand in 1993. Most codes written for S can still be used successfully in the R environment.

R is a powerful tool for analyzing data, plotting data or building statistical models for data. It also provides a wide variety of statistical and graphical techniques from basic to advanced and is greatly extensible. It complies and runs on various UNIX platforms and similar systems (i.e., Free BSD and Linux), Windows and MacOS. It is an ideal environment for statistical programming and visualization. R's four main advantages are described next.

Y. Chandra, L. Shang, *Qualitative Research Using R: A Systematic Approach*,
https://doi.org/10.1007/978-981-13-3170-1_4

Open Source R is free and open source, which means users can download and install it, and immediately start using it at *no cost*. Its authors make R source code available to others who wish to view and copy it, learn from it, modify it or share it. It has no license restrictions.

Statistical Programming R is unique because it enables users to *manipulate* (i.e., shape the dataset into a format that is easily accessed and analyzed by others), *analyze* (i.e., a wide variety of statistical analysis options, from basic to advanced) and *visualize* (i.e. various graph functions in R for data visualization) data in one single tool. The graphical capabilities of R are outstanding, and it provides a fully programmable graphics language.

R Packages The fundamental unit of shareable code in R is the package. R *packages* are collections of functions and data sets in a predefined format developed by the R community. *Extensibility* is one of the biggest advantages of R. Developers can easily write their own software and distribute it via *add-on packages*. Academics and data scientists commonly release their methods as packages, which means other users can directly access advanced and innovative methods. Currently, over 12,000 packages are freely available (https://cran.rproject.org/web/packages/available_packages_by_date.html).

R Community R has a thriving community that offers developers, collaborators and many other users a platform to discuss and share knowledge and expertise. It has active user groups (e.g., R-bloggers; Stack Overflow, R Conferences, GitHub) where questions can be asked and are quickly answered. The following Fig. 4.1 offers examples of R forums.

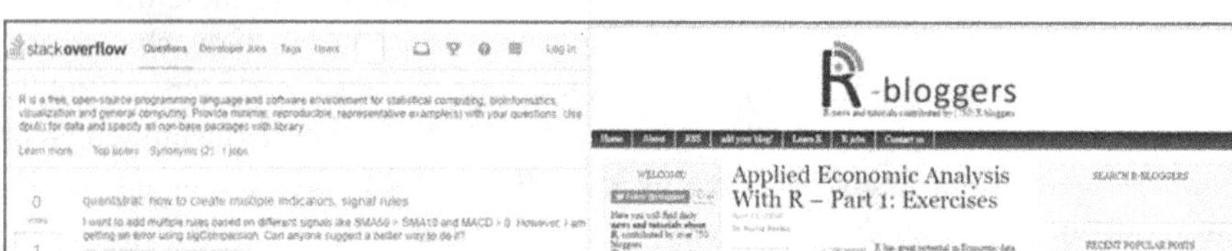

Fig. 4.1 A screenshot of R forum examples

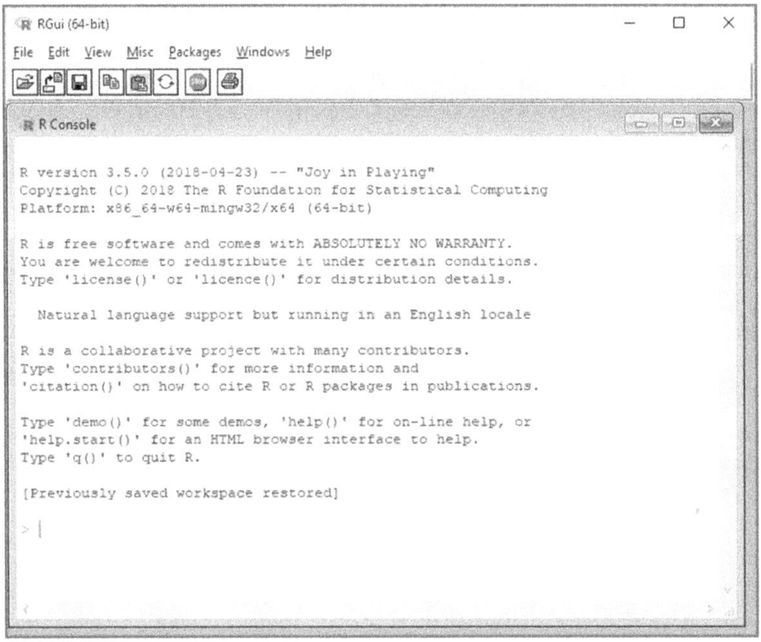

Fig. 4.2 A screenshot of R console

Writing syntax in R is easy. If you have used other languages such as C, C++, or Python before, R is easy and intuitive. Figure 4.2 illustrates how to write syntax in an R console (R Console is where computations are performed):

For more information on how to use R, go to R documentation at https://www.r-project.org/other-docs.html.

4.2 An Overview of RQDA

Given the large number of researchers who use R—an estimated two million users worldwide (Vance 2009), there are plenty of opportunities for R users to cross over from doing quantitative research to analyzing textual, qualitative data as a *primary* or *secondary* research tool. RQDA (Huang 2014) is an R package developed by Huang Ronggui for qualitative textual data analysis. As mentioned in the previous chapters, this book expounds upon the Gioia method and focuses on extending, elaborating upon and demonstrating this relatively new method of text analysis with the aid of RQDA as a new, open-source CAQDAS tool.

RQDA has some unique functions not easily found in the more popular CAQDAS tools (e.g., ATLAS.ti and NVivo) that can enhance the application of the Gioia approach to qualitative research. First, RQDA has built-in functionalities to perform *concept abstraction* from the first-order concepts to second-order themes. With additional creativity, by combining RQDA output with an Excel spreadsheet, a researcher can create third-order categories or *aggregate dimensions* (i.e., combining first-order and second-order codes to higher order codes) in a highly transparent manner to help readers and gatekeepers understand how the researcher made decisions in the data interpretation and aggregation process. Second, RQDA offers a functionality to *plot* the 'data structure' or a static picture that shows all important first-order and how they form second-order codes and aggregate dimensions. Third, RQDA is *open source* and thus *free* for any users who enjoy unlimited use. In the age of budget-cutting, universities, corporations and non-profits greatly benefit from free open-source software. Fourth, RQDA supports *data import into various languages*, including English and non-English documents such as Mandarin, Japanese, Korean, Arabic, German, French, Spanish, and Indonesian among other major languages. Fifth, RQDA is supported by SQLite (a relational database management system), which means that its functionality can be enhanced by coding with the R syntax. Because RQDA seamlessly integrates with R, it enables both statistical analysis of the coding and additional functions (e.g., visualization) for data manipulation and analysis. Thus, RQDA enables researchers to conduct *multi-method research* (combining qualitative and quantitative research) on a single platform. For example, Chandra and Shang (2017) used RQDA to qualitatively code the biographical profiles of social entrepreneurs and subsequently used the coded data to conduct a cluster analysis of the different biographical antecedents of social entrepreneurship. The last but not least unique feature of RQDA is that RQDA is a smaller part of R's open-source computing tools that have grown rapidly in both industry and academia (Culpepper and Aguinis 2011; Ihaka and Gentleman 1996; Vance 2009; Tippmann 2015). Therefore, access to RQDA instantly enables *access* to a larger, more powerful world of computing using the R language.

In the next chapter, we introduce how to download and install RQDA and to how to use it for qualitative data analysis with examples.

References

Chandra, Y., & Shang, L. (2017). Unpacking the biographical antecedents of the emergence of social enterprises: A narrative perspective. *Voluntas: International Journal of Voluntary and Nonprofit Organizations, 28*(6), 2498–2529.

Culpepper, S. A., & Aguinis, H. (2011). R is for revolution: A cutting-edge, free, open source statistical package. *Organizational Research Methods, 14*(4), 735–740.

Huang, R. G. (2014). *RQDA: R-based qualitative data analysis. R package version 0.2-7.* Retrieved from http://rqda.r-forge.r-project.org. on 18 Apr 2018

Ihaka, R., & Gentleman, R. (1996). R: A language for data analysis and graphics. *Journal of Computational and Graphical Statistics, 5*(3), 299–314.

Tippmann, S. (2015). Programming tools: Adventures with R. *Nature News, 517*(7532), 109.

Vance, A. (2009). Data analysts captivated by R's power. *New York Times*, *6*(5.4). Retrieved from http://www.nytimes.com/2009/01/07/technology/business-computing/07program.html?pagewanted=all&_r=0. on 18 Apr 2018.

Chapter 5
Installing RQDA

Abstract As an R package, RQDA can be easily installed in R base or RStudio if you are familiar with the basics of the R programming language. By accessing the R base, RQDA can be installed in different operating systems, including Windows, Linux, and Mac OSX. Windows users can use an alternative route to install RQDA without accessing R (as we will discuss in this chapter). RQDA is a user-friendly tool for both R users and non-R users because it has a graphical user interface (GUI). That is, users can 'point and click' just like using Windows software. As noted in the previous chapter, RQDA and its supporting software programs (e.g., RSQLite for data storage, Gtk+ for GUI) covered in this book are *open source*, meaning that users can download them for free. In this chapter, we discuss how to install RQDA using syntax for R users and R novices.

Keywords RQDA installation · RQDA package · RStudio · R · CRAN · R syntax

5.1 For R Users

RQDA is available for Windows, Mac, and Linux operating systems and can be accessed through an R platform: R base or RStudio. To install RQDA in Windows and Linux OS, follow the three steps shown in Fig. 5.1.

Step 1: Install GTK+

As the first step, before RQDA can be installed, users must first install GTK+, a toolkit for creating graphical user interfaces (GUI) because RQDA is based on RGtk2, which depends on GTK+. To make this installation, Windows users can click the following link for SourceForge:

https://sourceforge.net/projects/gladewin32/postdownload?source=dlp

or via this path below.

https://sourceforge.net/projects/gladewin32/files/gtk%2B-win32-runtime/2.10.11/gtk-2.10.11-win32-1.exe/download?use_mirror=ncu&modtime=1175123376&big_mirror=0

Y. Chandra, L. Shang, *Qualitative Research Using R: A Systematic Approach*,
https://doi.org/10.1007/978-981-13-3170-1_5

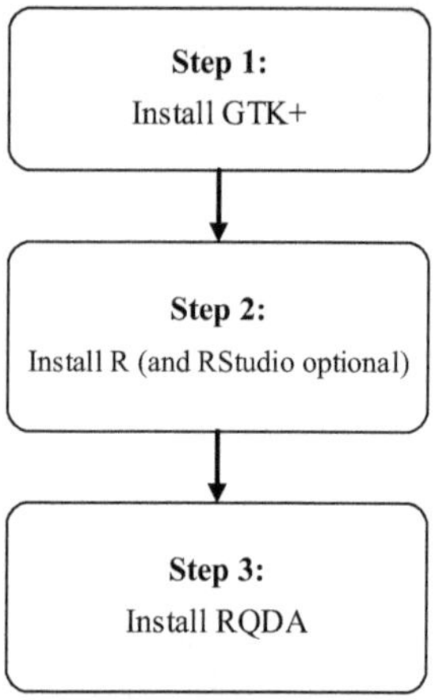

Fig. 5.1 Installing RQDA for Windows and Linux users

to download and install GTK+. After you enter the website, the download will automatically start within a few seconds.

For Debian-based (including Sid-based, Testing-based and Stable-based) Linux users, to install GTK+, open a 'Terminal' inside the graphical environment and type *sudo apt-get install libgtk2.0-dev.* For Fedora users, type in *yum install -y gtk2-devel* to install.

Other *Linux users* may need to build and install GTK+ manually. To do so, first install necessary external dependencies, which include pkg-config, GNU make, the JPEG, PNG, and TIFF libraries, FreeType, and, if necessary, libiconv and libintl.

Next, install the GTK+ libraries in this order: Glib, Pango, ATK, and GTK+. For more information on installing GTK+ manually in a Linux operating system, see this website: https://developer.gnome.org/gtk3/unstable/gtk-building.html.

Step 2: Install R Base (or RStudio)

After GTK+ is successfully installed on your computer, download and install R. R base (R project) and RStudio (optional) must be installed before installing RQDA because RQDA runs on the R platform. To install R base, go to CRAN: https://cran.r-project.org/, and you should see a page as shown in Fig. 5.2:

CRAN is a network of ftp (file transfer protocols) and web servers around the world that store different versions of code and documentation for R. The links in this website enable you to download the same software. For faster download speed, click a location that is geographically *closest* to you. After this, you should see a page with various locations that host R (e.g., London, Shanghai, or Tokyo); choose one that you want to download and install (as shown in Fig. 5.3).

Please select the version that you want to download based on the operating system that you are using. For example, for Windows users, please select *Download R for Windows*. After this, you will see a page with prompts to choose which subdirec-

Fig. 5.2 Screenshot of the CRAN mirrors website. (https://cran.r-project.org/)

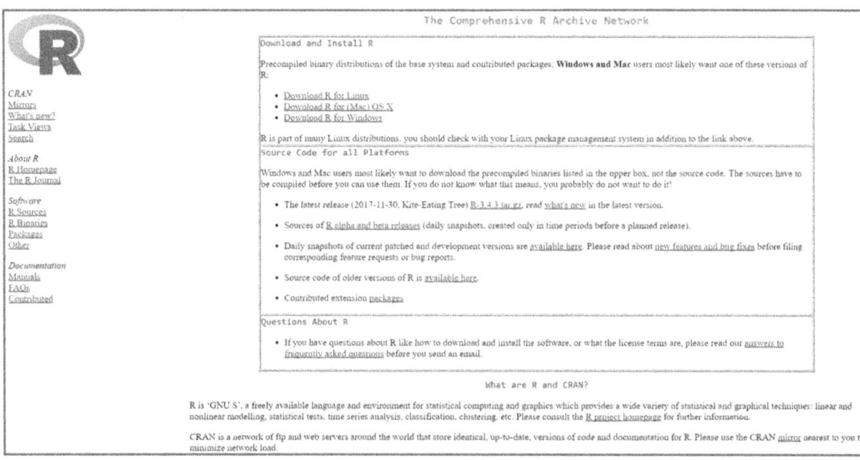

Fig. 5.3 Downloading and installing R. (https://cran.r-project.org/)

tories (e.g., your "Desktop" or a particular folder in your computer) you want to download and install as shown in following Fig. 5.4.

On this page, click *install R for the first time* and this will take you to the download page. The suggested version of R in this page supports both 32-bit and 64-bit Windows (check if your desktop is 32- or 64- bit and then download the correct version of R). If you want to know more about downloading R in a Windows, see the FAQ page: https://cran.r-project.org/doc/FAQ/R-FAQ.html.

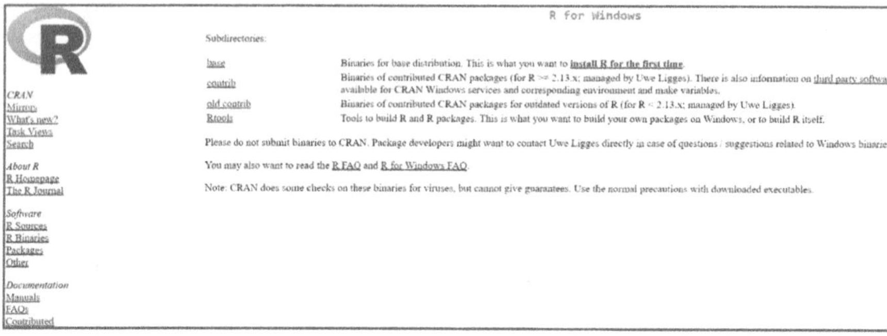

Fig. 5.4 Subdirectories for Windows users. (https://cran.r-project.org/)

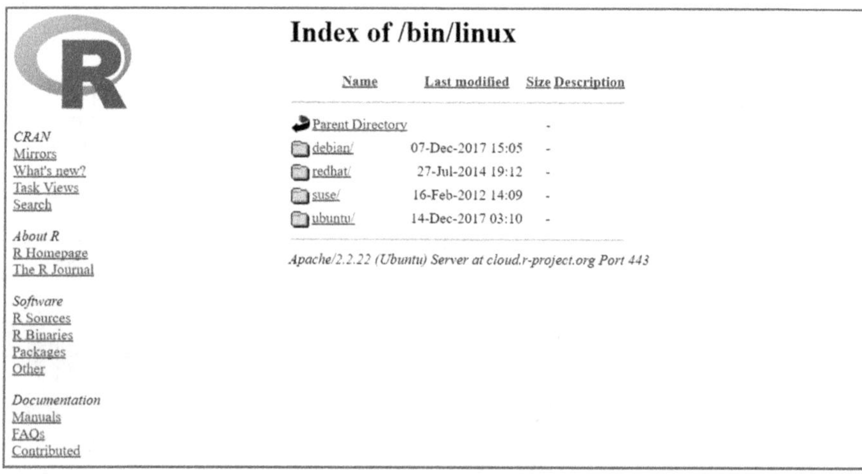

Fig. 5.5 Downloading R for a Linux-based operation system (https://cran.r-project.org/)

For Linux users, to download and install R, click *Download R for Linux* and you will see the page shown in Fig. 5.5:

To get instructions for installing R, choose the Linux version you are using. For example, if you are using a Debian-based Linux, choose *ubuntu/*, which will take you to the page shown in Fig. 5.6. (see https://cran.r-project.org/").

Instructions for installing R can be found by clicking *Installation* under *UBUNTU PACKAGES FOR R* located where on the page shown in Fig. 5.6 below. For instance, to install R on a Ubundu-based Linux operating system, open a Terminal and type *sudo apt-get install r-base* as shown in Fig. 5.7.

After successfully downloading and installing R base on your computer, you can install *RStudio,* but this is optional. RStudio is a free open-source integrated development environment (IDE) for R that provides comprehensive tools for software

Fig. 5.6 Downloading R for a Ubuntu-based Linux OS

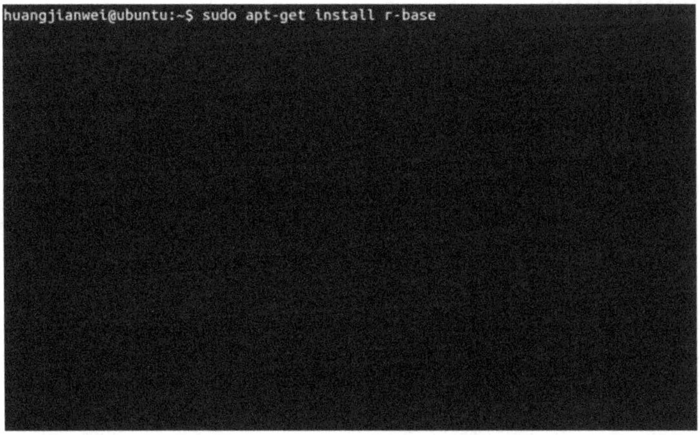

Fig. 5.7 Installing R in a Ubuntu-based Linux OS

development. RStudio makes it easier to do R coding and is very useful for exploratory data analysis. For example, RStudio users can seamlessly document what they are doing and conveniently save figures and outputs. The Fig. 5.8 screenshot shows the RStudio interface. In the console window in the bottom left panel, type a syntax (to give instructions to the computer to perform tasks). In RStudio, you can type any syntax directly into the console, or you can enter any syntax in the script editor and save the script in case you forget the syntax when you close the session. In this way, you can track what you did with the syntax and can easily share your syntax with others.

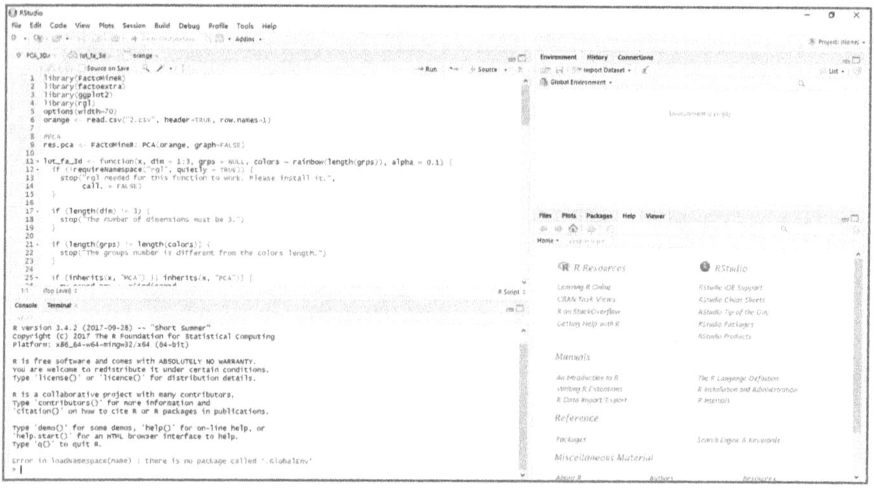

Fig. 5.8 RStudio user interface

Step 3: Install RQDA

The last step is to install RQDA on the R platform (or alternatively on RStudio). For Windows and Linux users (Mac instructions follow), to install RQDA you must launch R (or RStudio) and type the following syntax in the R Console:

```
install.packages("RQDA",dependencies=c("Depends", "Imports"))
```

The Fig. 5.9 screenshot shows how to install RQDA by typing the package installation syntax in RQDA (see bottom left of Fig. 5.9):

After typing a syntax in the console window, press *enter*, and the download and installation of RQDA package should start. If you prefer to copy and paste the above syntax to R, first copy the syntax to a notepad (available in any Windows based computer) to eliminate any programming codes that could be embedded in Microsoft Word. From the notepad, copy and paste the syntax to R (or RStudio). If the above step is successful, you can then launch RQDA by typing the syntax *library(RQDA)* from within R base (or RStudio), and the RQDA panel will open in a separate window (see Fig. 5.10). The screenshot in Fig. 5.10 shows that the RQDA software is ready to use.

For Mac OS users, the installation process of RQDA is slightly different. Figure 5.11 summarizes the four steps to install RQDA in a Mac OS:

If you are using a Mac OS, you need to first install X11, a windowing system for bitmap displays, which provides the basic framework for a GUI environment.

To download X11, you can go to https://www.xquartz.org/, and you will see a page similar to that shown in Fig. 5.12:

You can click *XQuartz-x.x.x.dmg* (x.x.x. stands for the software version numbers) and the download will start within a few seconds. Next, check the *Info* section

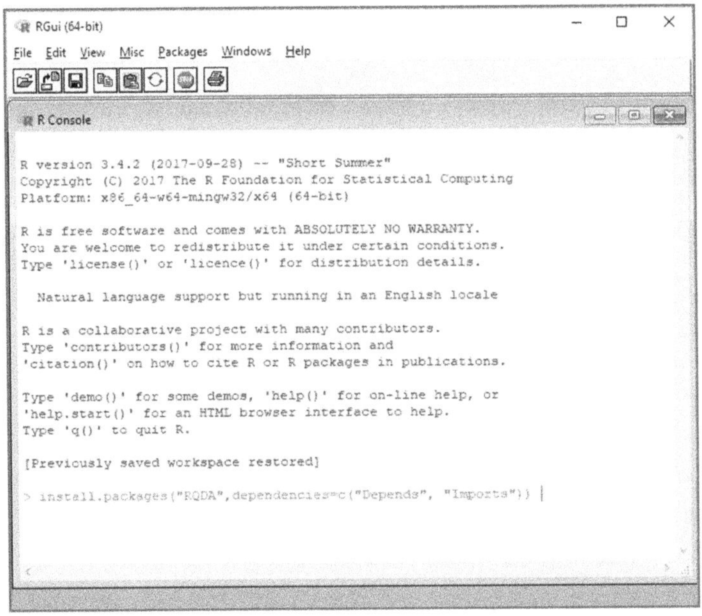

Fig. 5.9 Inputting a syntax to install RQDA in R

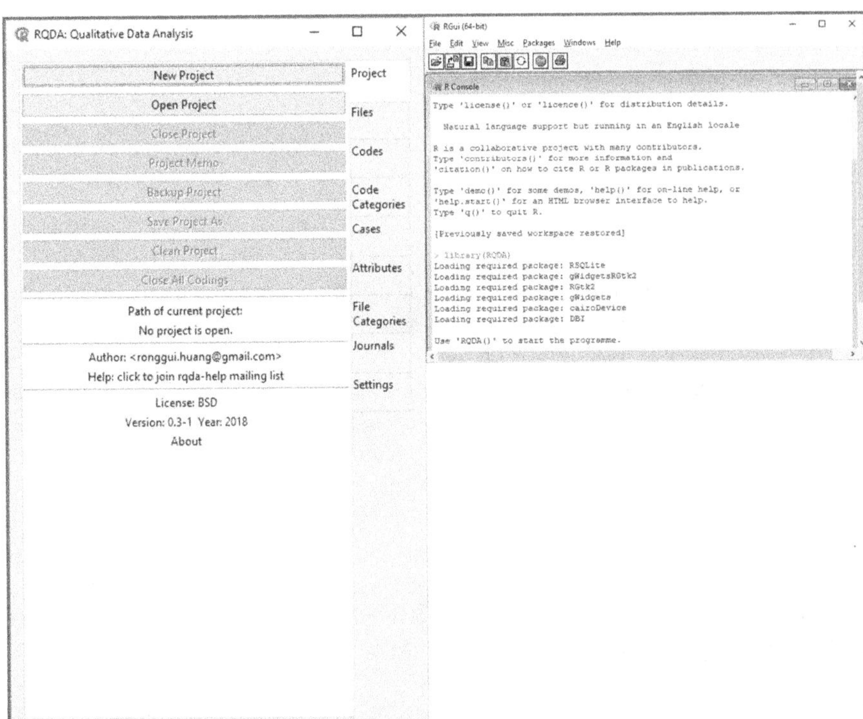

Fig. 5.10 RQDA panel launched from R

Fig. 5.11 Installing RQDA for Mac OS users

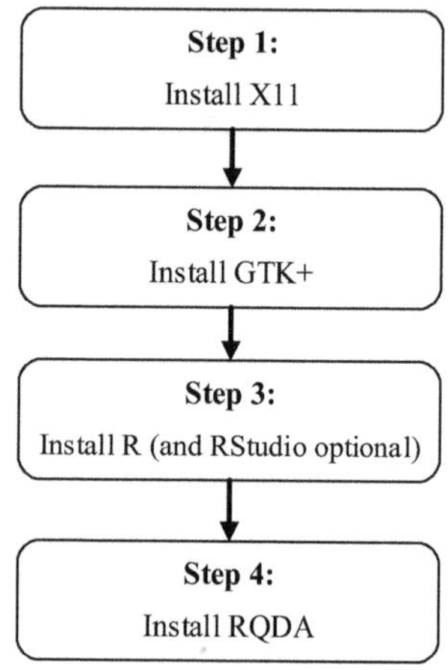

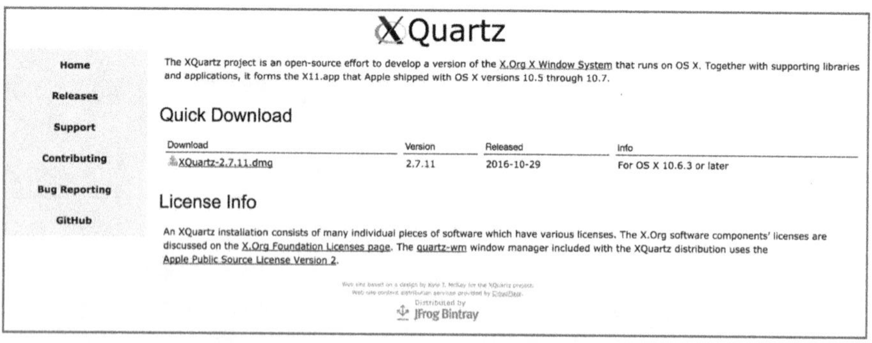

Fig. 5.12 X11 downloading page. (https://www.xquartz.org/)

to see whether your operating system accepts the newest version of XQuartz. If not, go to *Releases* on the left toolbar to find previous versions of XQuartz.

After successfully installing X11 on your Mac OS, the next step is to install GTK+. To do so, go to http://r.research.att.com and look for a page of R for Mac OS X developers. Please click on the '*Other binaries and tools (e.g., RSwitch, Graphviz, GTK+,...)*' under Index (as shown in Fig. 5.13), and find a hyperlink for downloading the GTK+ 2.24.17 framework from the *Other binaries* section.

After this, go to the CRAN download page: https://cran.r-project.org/bin/macosx/ and download and install R for Mac OS, and you will see the page below (Fig. 5.14):

The latest version of R (17 December 2017) is R-3.5.0. Note that R-3.5.0 supports only Mac OS X release 10.6 and newer. If you have an older version of Mac OS, find the last supported release of R from the page at https://cran.r-project.org/bin/macosx/old/. For users who also want to download and install RStudio on a Mac

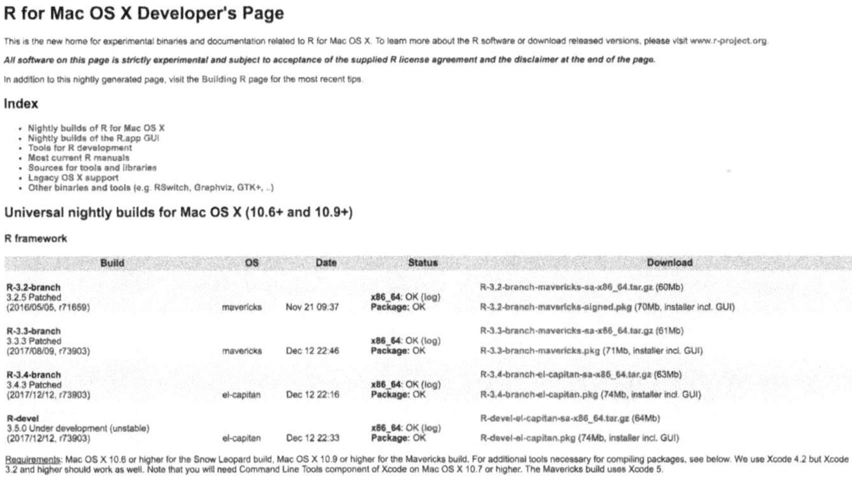

Fig. 5.13 A screenshot of R for Mac OS X developer's page. (https://r.research.att.com/)

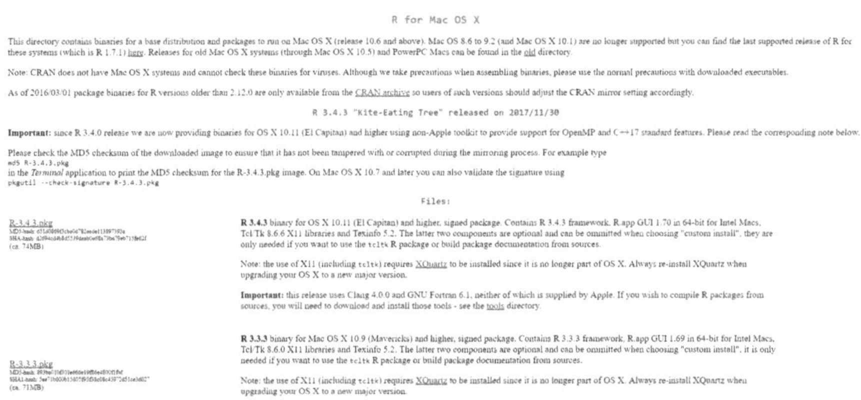

Fig. 5.14 Screenshot of download page for R for Mac OS

OS, go directly to the RStudio download page: https://www.rstudio.com/products/rstudio/download/#download and select the version that matches your operating system (e.g., Mac OS X 10.6).

After you have successfully installed R on your Mac OS, next install RQDA from within the R platform by typing in this syntax:

```
install.packages('RQDA')
```

If all steps are successful, you can launch RQDA using syntax *library (RQDA)* from within R or RStudio.

5.2 For Users New to R

The installation of RQDA can be quite complicated for users who have no experience with the R programming language because it involves a bit of programming, which requires the installation of GTK+ and R. However, the RQDA developer (Huang 2014) provides an alternative route for non-R users to install RQDA *without* accessing the R platform. Thus, users who use a Windows-based operating system can download the RQDA.bat file from OneDrive. Users can click https://onedrive.live.com/?id=41C1EE655922D297%21278&cid=41C1EE655922D297 and choose a suitable version of RQDA (e.g., RQDA 0.2-5.rar). The RQDA file will then begin to download. Next, users can launch RQDA by double-clicking the RQDA.bat file. A separate R console window will then appear (see Fig. 5.15).

To launch RQDA, click *Packages* from the toolbar and select *Load package*, and a new window will open with names of different R packages as follows (see Fig. 5.16). Choose *RQDA* from the menu and click *OK*, and RQDA should load automatically in the R console.

The R is very sensitive to imperfect syntax typed in to the console. For example, typing a syntax using 'upper case' where it requires a lower case will create an 'error'. Another type of error is when you add an 'indented space' or 'unnecessary spacing' in the syntax where it is not needed.

Now that you have successfully imported the RQDA and R programs, you can now begin preparing your data sampling and importing process.

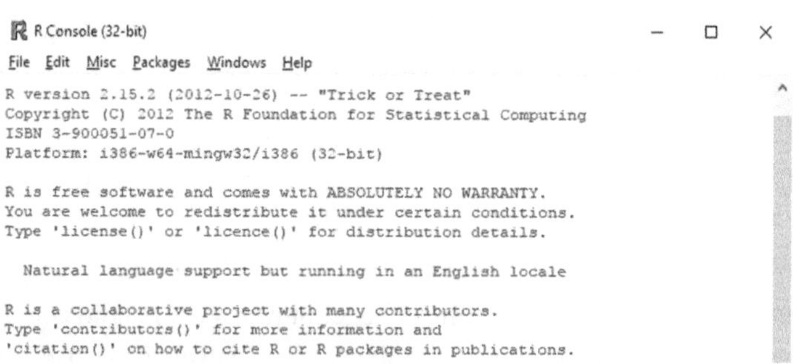

Fig. 5.15 R console launched from RQDA.bat

Fig. 5.16 A pop-up window from RQDA

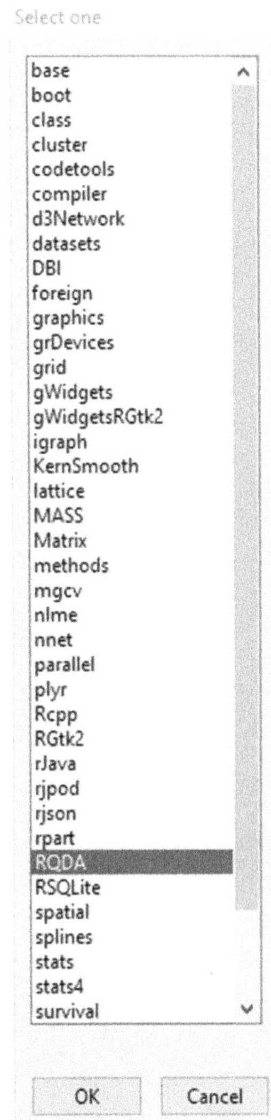

Chapter 6
Data Preparation and Sampling

Abstract After installing RQDA, the next step is to import qualitative data that you want to analyze. This chapter discusses the techniques to convert qualitative data (i.e., words) into the data format supported by RQDA. We also offer techniques to combine several pieces of files into a corpus (a corpus file) and other "tricks" to prepare the data for RQDA analysis. With respect to primary data collection, it discusses how to systematically conduct sampling (e.g., for interviews, news articles).

Keywords Social innovation · Ashoka Changemakers · Sampling · Data collection · Data preparation

6.1 Sampling

Social Innovation Strategy and Outcomes

To help illustrate this process, we examine a representative project to demonstrate, step by step, the sampling process and data preparation. Our representative project seeks to understand "the impacts brought about by social innovation (SI) in tackling social problems around the world". This project is used to demonstrate the use of Gioia approach to qualitative research with the aid of RQDA. This model project employs SI, a relatively new approach to solving social problems that is more effective, efficient and sustainable than existing solutions (e.g., government programs) (Nicholls et al. 2015; Phillips et al. 2015). SI is seen as an important approach to solving many social problems ranging from chronic poverty to lack of healthcare, education, food, and electricity. SI develops social practices created from collective, intentional, goal-oriented actions aimed at promoting social change (Cajaiba-Santana 2014). It generates and implements innovative grassroots projects that involve community members, social organizations, social relationships, and social institutions (Mumford 2002) in an effort to solve social problems.

Y. Chandra, L. Shang, *Qualitative Research Using R: A Systematic Approach*,
https://doi.org/10.1007/978-981-13-3170-1_6

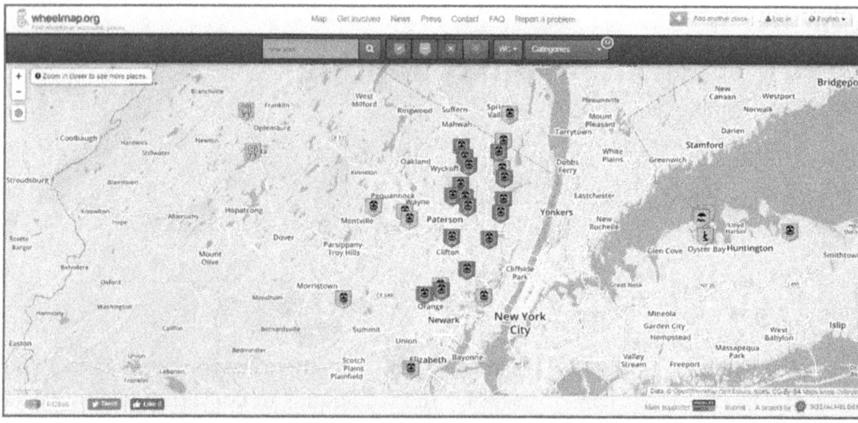

Fig. 6.1 A screenshot of Wheelmap.org

For example, Wheelmap.org, an SI project developed by a Berlin, Germany-based nonprofit Sozialhelden (Social Heroes), provides online maps to assist handicapped individuals to find shops, subways, movie theatres, et cetera, accessible via wheelchair (see Fig. 6.1). Currently, over 750,000 public places worldwide can be found on the map and it helps thousands of people with disabilities navigate their worlds.

Despite the growing interest in social innovation, there is little academic research that systematically explores strategies and social impacts of SI projects in solving complex health challenges. We still know little about what strategies are used by social innovators, how social innovators make healthcare services accessible to disadvantaged (e.g., poor, handicapped, rural) populations, in what ways these innovators have brought changes to patients, healthcare providers and even institutions and so forth. Therefore, our illustrative project asks two important yet unanswered questions: (1) *How have social innovators solved healthcare challenges?* (2) *How can social innovators create social impacts in the healthcare sector?*

To answer these questions, we selected SI samples from a Social Innovation competition hosted by Ashoka Changemakers called *Making More Health: Co-Creating a Healthier World*, one of the world's largest social innovation 'tournaments' (Terwiesch and Ulrich 2009). More than 15,000 innovators with over 20,000 SI projects worldwide have submitted proposals for this program from early 2000 to January 2018. Ashoka is a network of social innovators and impact partners working collaboratively with a purpose to accelerate change for critical social problems.

The focal competition *"Making More Health: Co-Creating a Healthier World"* (https://www.changemakers.com/makingmorehealth?qt-competition_info_section=1&qt-competition_entries=zero#qt-competition_entries), or "MMH", was launched on 16 September 2015. It sought to attract social innovations with a person-centered approach to health and with the commercial potential to scale up through a partnership with Boehringer Ingelheim (a global pharmaceutical company). A total of 157 entries from over 40 countries competed and 47 were 'semifinalists, finalists, and the winners'. Each of the entries contains a *profile* of the social innovation: their goals, their strategies to solve social problems, their impact, and information about the social innovators' background. The screenshot in Fig. 6.2 offers a glimpse into the competition.

In our illustrative project, we used a *stratified sampling technique*[1] and selected 10 project profiles consisting two distinct groups—five "semifinalists" (n = 47) and five "non-winners" (n = 110)—for analysis. Therefore, we sampled for the dependent variable (whether a social innovation reached a semifinal stage or failed to reach semifinal stage) and coded the SI profiles to derive codes and patterns that explain the strategies and social impact created by the social innovators. Because the research questions are about "how the social innovators solve social problems" and "their outcomes", our data coding will focus on words, phrases, and sentences that reflect the solutions employed by the social innovators and how the solutions create outcomes. Table 6.1 shows our sampling process:

Firstly, we extracted all 10 project profiles from the Ashoka Changemakers website. Figure 6.3 shows a sample SI project profile.

In our illustrative project, we triangulated the SI profile data by collecting various types of data from various sources (e.g., news articles of each social innovator, social media posts from the social innovator, their personal/organizational websites) to explore the strategies and impacts of social innovations in the healthcare sector. Additional biographical information from non-Ashoka Changemaker sources (e.g., media interviews, news articles about the competition and the Changemakers projects) online was also collected. Table 6.2 details the 10 social innovation project names along with the additional data sources used.

[1] Given the availability of social innovation database such as Ashoka Changemakers and the relatively large size of the data of social innovation projects, we did employ a purposeful sampling. Rather, we used a stratified sampling approach of the database to enhance the model study's robustness.

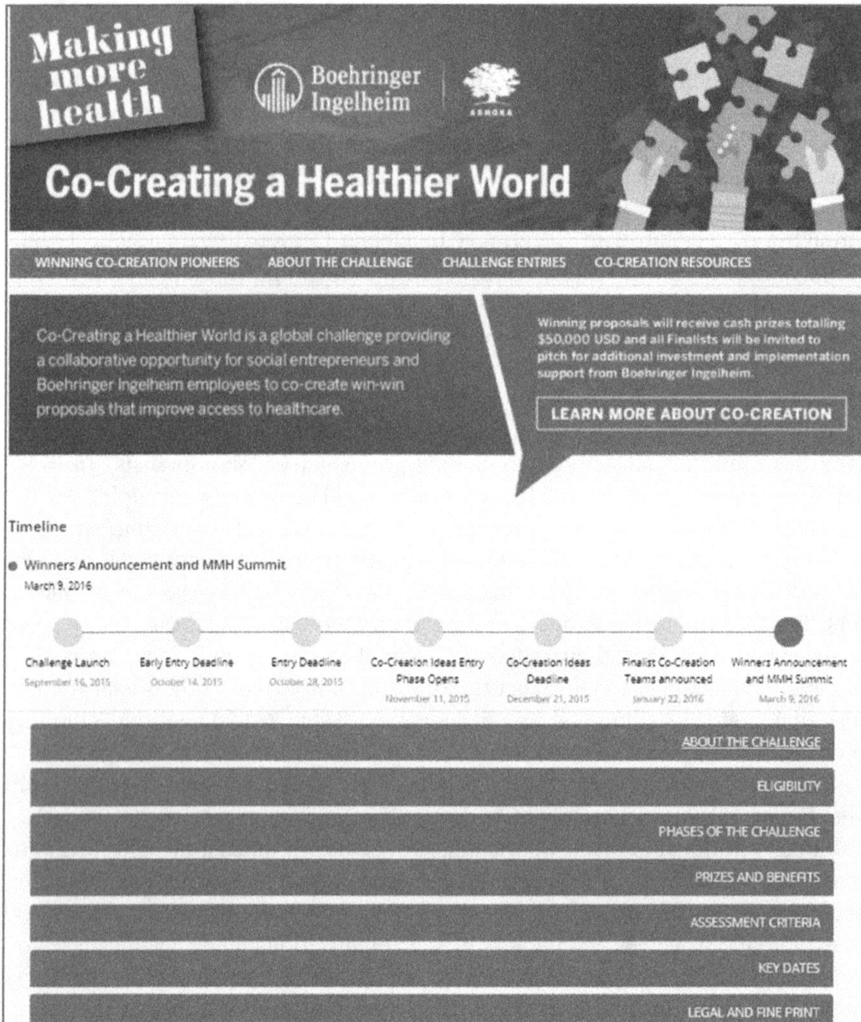

Fig. 6.2 A screenshot of the MMH website

Table 6.1 Sampling for the illustrative project

Making more health: co-creating a healthier world	Population		Samples	
Semi-finalists (including winners, finalists & semi-finalists)	47	30%	3	30%
Non-winners	117	70%	7	70%
Total	164	100%	10	100%

BAS VAN ABEL

INTRODUCTION / CITATION / MORE FOR YOU

ASHOKA FELLOW

SHARE

NETHERLANDS

FELLOW SINCE 2016

FAIRPHONE

INTRODUCTION

By designing and producing the first modular designed. and thus longer lasting, smartphone, building a conflict free and fair supply and production chain. Bas van Abel with his company Fairphone is creating a movement of ethical consumption while shifting the smartphone industry -one of the most complex supply chains- towards ethical production and complete transparency, fueling change instead of conflict.

THE NEW IDEA

Bas founded Fairphone to build a movement for fairer electronics, using three strategies; Firstly, by creating a smartphone which shows the whole electronics industry it is possible to produce fair and sustainable products even in one of the most challenging and complex industries (electronic consumer goods). Secondly, by designing for re-use and longer life-cycles of the various elements of the phone, he introduced the first modular phone on the market. And thirdly, by 'raising' a new generation of consumers that is aware of the way electronic products are designed and produced, literally creating a market for electronic products that put ethical values first. By doing this Fairphone creates a social innovation platform for all stakeholders in the electronics supply chain.

THE STRATEGY

Bas van Abel founded Fairphone in 2013 to build a movement that would disrupt one of the most challenging, complex and massive industries around: the smartphone industry. Bas is motivated by the philosophical perspective of our capitalist system and solving the challenges it brings with it. By showing that fair and sustainable design, production and consumption is possible in one of the dirtiest industries, Bas shows the profit-driven system is failing us and that we can build a better alternative. Ultimately he strives for a human centered system. Bas envisions a new economy – one where social and environmental values become a natural part of doing big business.

Bas is revolutionizing the consumer electronics and smartphone industry by focusing on three key pillars.
First, he positively influences key players in the supply chain. For example, Fairphone has ensured improved safety and labor measures with their manufacturing contractor in Chongqing, China. These include improved fire safety measures, written policies on child labor and juvenile protection, proper and adequate labelling of chemicals, improved lighting in the production rooms, free lunch in the cafeteria and minimum wage plus a premium which goes into a Worker Welfare Fund. Through these efforts, more than $300k in premiums was paid out to workers through this worker welfare fund for the entire production facility (even though Fairphone workers only make up a small part). All these changes have been realized by a democratically elected workers council, a first in China where freedom of association is severely restricted. Fairphone therefore creates a change in manufacturing practices aiming to create spillover effects in the supply chain to other smartphone manufacturers (that buy from this production facility); simply by showing to other big electronic manufacturers that more sustainable and fair manufacturing practices are possible.

With regards to the mining of minerals and metals, Fairphone chooses and even sets up certified mines wherever possible and pushes systematically for better mining practices. Together with the Fairtrade organization and a Peruvian gold mine, they have recently successfully set up the world's first Fairtrade gold mine, a mine of large capacity with clients such as Apple and Samsung. Through this specific project, producers beyond Fairphone are now getting access to Fairtrade gold. They have impacted this part of the supply chain in such a way that other large players (like Samsung and Apple) are now are able to source Fairtrade gold too, increasing impact beyond Fairphone's own production.

Other achievements illustrating how Bas is building a fairer electronics manufacturing supply chain, include the setting up of a conflict-free tungsten supply chain through several conflict-free mining programs in the Democratic Republic of Congo, which were catalysts in increasing

Fig. 6.3 A sample profile of social innovators from Ashoka Changemakers

Table 6.2 Data source used in this demonstration

	Data source					
SI Projects	Ashoka ChangeMakers website	Other fellowship/ award websites	Organization official websites	Online articles	News	Interviews
MicroClinic Technologies	X	X	X		X	X
Changmka Microhealth	X		X	X	X	
Healthy Factories	X		X		X	
Apart of Me	X	X	X	X		X
Project Oxygen	X	X		X		
FullSoul Health Solutions	X		X			
Drink to Life	X					
EyeCheck Solutions	X		X	X		
The Green Journey	X		X			
Mobile Surgery	X		X			

6.2 Data Preparation

After collecting all relevant data (as we did in Table 6.2 above), we then combined all the data for each project and saved it as a "single file" for each SI project. Thus, we created 10 separate files for the 10 projects (or cases) in the demonstration. RQDA only supports documents in "plain text" format. Therefore, before we imported the data files into RQDA, we saved the data in plain text format (*.txt*). Please note data import to RQDA requires *individual files* (instead of one large combined or corpus file). (Therefore, we suggest that you save your data into separate text files and label them accordingly for ease of analysis in RQDA.) In our sample project, the file is named "Organization Name_Semifinalists/Nonwinner". This will allow us to easily identify a coded data and the organization it is associated with. Figure 6.4 shows an actual sample of a project profile in .txt format.

If your data files are not in .txt format, convert them into plain text format. This can be done by copying a body of textual data (e.g., an interview transcript or profiles of a social innovator) to a Word file (a data file, either a .doc or docx is fine) and then save as *a* .txt file. RQDA also has a function to help users convert portable document format (PDF) files to plain text (.txt) files by typing R syntax. To do this, users must first download pdftotext.exe from http://www.foolabs.com/xpdf/download.html. Next, type a few lines of syntax into the R console. Then create a folder

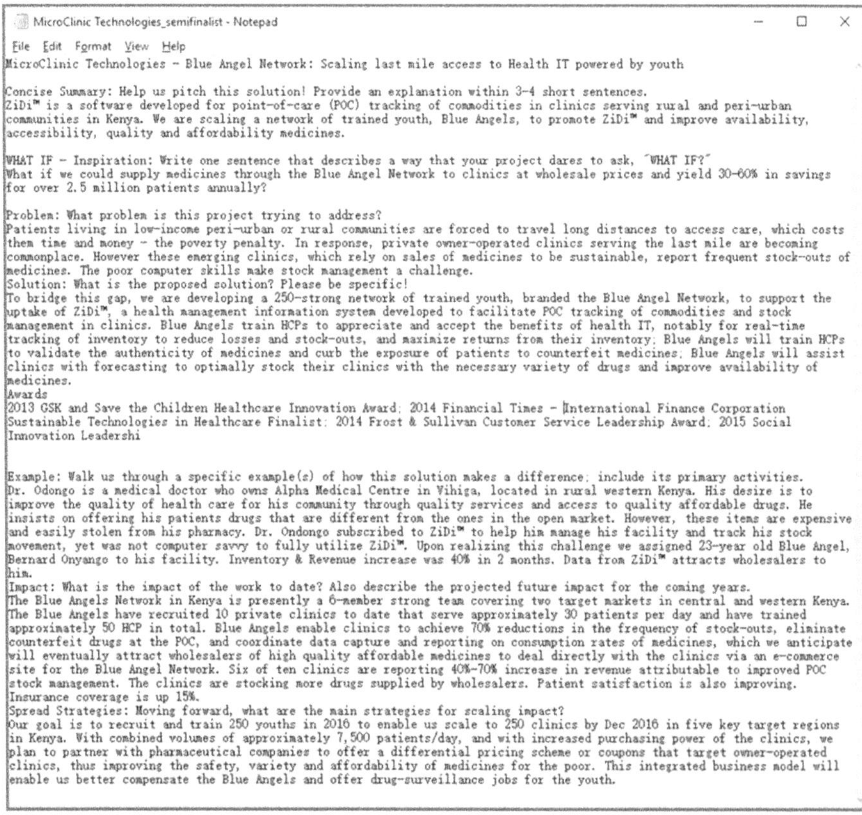

MicroClinic Technologies_semifinalist - Notepad

File Edit Format View Help

MicroClinic Technologies - Blue Angel Network: Scaling last mile access to Health IT powered by youth

Concise Summary: Help us pitch this solution! Provide an explanation within 3-4 short sentences.
ZiDi™ is a software developed for point-of-care (POC) tracking of commodities in clinics serving rural and peri-urban communities in Kenya. We are scaling a network of trained youth, Blue Angels, to promote ZiDi™ and improve availability, accessibility, quality and affordability medicines.

WHAT IF - Inspiration: Write one sentence that describes a way that your project dares to ask, "WHAT IF?"
What if we could supply medicines through the Blue Angel Network to clinics at wholesale prices and yield 30-60% in savings for over 2.5 million patients annually?

Problem: What problem is this project trying to address?
Patients living in low-income peri-urban or rural communities are forced to travel long distances to access care, which costs them time and money - the poverty penalty. In response, private owner-operated clinics serving the last mile are becoming commonplace. However these emerging clinics, which rely on sales of medicines to be sustainable, report frequent stock-outs of medicines. The poor computer skills make stock management a challenge.
Solution: What is the proposed solution? Please be specific!
To bridge this gap, we are developing a 250-strong network of trained youth, branded the Blue Angel Network, to support the uptake of ZiDi™, a health management information system developed to facilitate POC tracking of commodities and stock management in clinics. Blue Angels train HCPs to appreciate and accept the benefits of health IT, notably for real-time tracking of inventory to reduce losses and stock-outs, and maximize returns from their inventory; Blue Angels will train HCPs to validate the authenticity of medicines and curb the exposure of patients to counterfeit medicines; Blue Angels will assist clinics with forecasting to optimally stock their clinics with the necessary variety of drugs and improve availability of medicines.
Awards
2013 GSK and Save the Children Healthcare Innovation Award; 2014 Financial Times - International Finance Corporation Sustainable Technologies in Healthcare Finalist; 2014 Frost & Sullivan Customer Service Leadership Award; 2015 Social Innovation Leadershi

Example: Walk us through a specific example(s) of how this solution makes a difference; include its primary activities.
Dr. Odongo is a medical doctor who owns Alpha Medical Centre in Vihiga, located in rural western Kenya. His desire is to improve the quality of health care for his community through quality services and access to quality affordable drugs. He insists on offering his patients drugs that are different from the ones in the open market. However, these items are expensive and easily stolen from his pharmacy. Dr. Ondongo subscribed to ZiDi™ to help him manage his facility and track his stock movement, yet was not computer savvy to fully utilize ZiDi™. Upon realizing this challenge we assigned 23-year old Blue Angel, Bernard Onyango to his facility. Inventory & Revenue increase was 40% in 2 months. Data from ZiDi™ attracts wholesalers to him.
Impact: What is the impact of the work to date? Also describe the projected future impact for the coming years.
The Blue Angels Network in Kenya is presently a 6-member strong team covering two target markets in central and western Kenya. The Blue Angels have recruited 10 private clinics to date that serve approximately 30 patients per day and have trained approximately 50 HCP in total. Blue Angels enable clinics to achieve 70% reductions in the frequency of stock-outs, eliminate counterfeit drugs at the POC, and coordinate data capture and reporting on consumption rates of medicines, which we anticipate will eventually attract wholesalers of high quality affordable medicines to deal directly with the clinics via an e-commerce site for the Blue Angel Network. Six of ten clinics are reporting 40%-70% increase in revenue attributable to improved POC stock management. The clinics are stocking more drugs supplied by wholesalers. Patient satisfaction is also improving. Insurance coverage is up 15%.
Spread Strategies: Moving forward, what are the main strategies for scaling impact?
Our goal is to recruit and train 250 youths in 2016 to enable us scale to 250 clinics by Dec 2016 in five key target regions in Kenya. With combined volumes of approximately 7,500 patients/day, and with increased purchasing power of the clinics, we plan to partner with pharmaceutical companies to offer a differential pricing scheme or coupons that target owner-operated clinics, thus improving the safety, variety and affordability of medicines for the poor. This integrated business model will enable us better compensate the Blue Angels and offer drug-surveillance jobs for the youth.

Fig. 6.4 A sample social innovation project text file

with ten PDF files to be converted into text files and to import them into RQDA for qualitative data analysis, type the following syntax (first copy and paste the syntax to a 'notepad' and then copy it into the R console; In R, the pound sign denotes a comment line, and everything to the right of # in a line is ignored by the interpreter):

```
dest <- "C:\\Users\\Desktop" #location of the folder that contains 10 pdf files#
myfiles <- list.files(path = dest, pattern = "pdf", full.names = TRUE)
lapply(myfiles,function(i)system(paste('"C:/ProgramFiles/xpdf/bin64/pdftotext.exe"', paste0('"', i, '"')), wait = FALSE) ) #location of pdftotext.exe#
```

Imagine if you have hundreds or more PDF files. With the syntax above, you can save a lot of time converting them to .txt files and importing them into RQDA. Figure 6.5 shows a screenshot of the syntax we wrote above in RStudio:

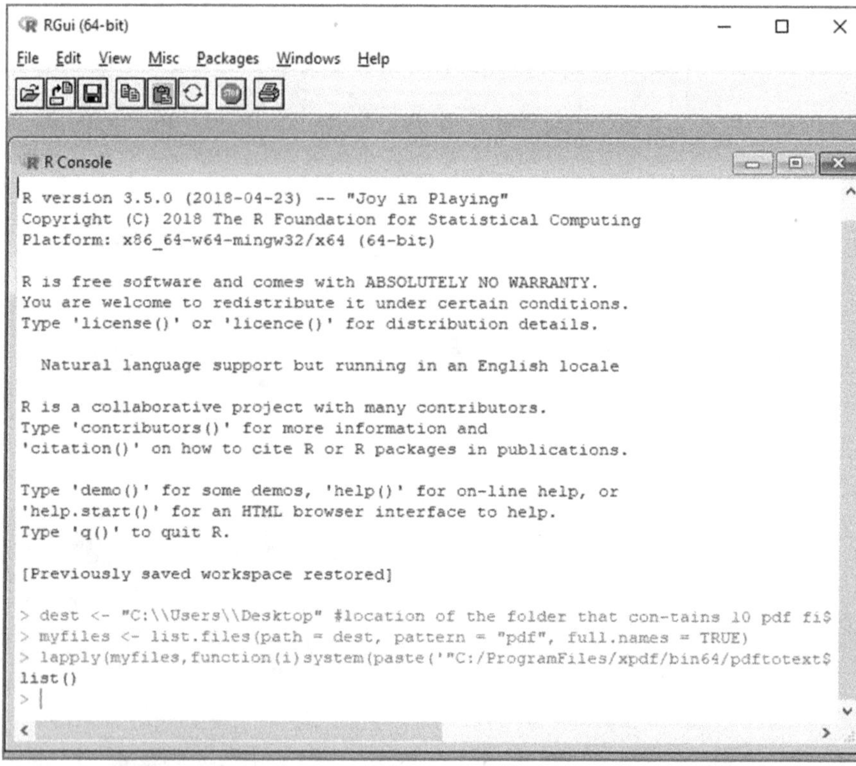

Fig. 6.5 Converting PDF files into text files in R

For users who do not want to use or write syntax in R, open source (free) applications can be used to convert a batch of files to .txt files. For example, Convert Doc, an easy-to-use document converter for Windows users can be downloaded from http://www.softinterface.com/Convert-Doc/Convert-Doc.htm as shown in Fig. 6.6.

Additionally, you can use a built-in Microsoft Office function to convert your documents into the .txt format. For Mac users, there is a tool called *Textutil* in Mac that enables quick .txt file format conversions. Users can go to

http://osxdaily.com/2016/11/05/convert-pdf-to-docx-mac-os-x/

and follow the instructions to convert files to text files

In addition to the file type requirement (i.e., plain text (.txt) format), RQDA also enables users to import highlighted texts (see Fig. 6.7 for an example of PDF highlights) in PDF files. Users often used a highlighter to color texts inside a PDF document and these *highlighted texts* can be captured in RQDA. Therefore, an alternative way to prepare your data is to import them from PDF files using the *import highlights* function in RQDA.

To import PDF highlights into RQDA, an appropriate PDF reader such as Adobe Acrobat Pro, Foxit Reader, Nitro PDF Reader, Slim PDF, Expert PDF Reader or

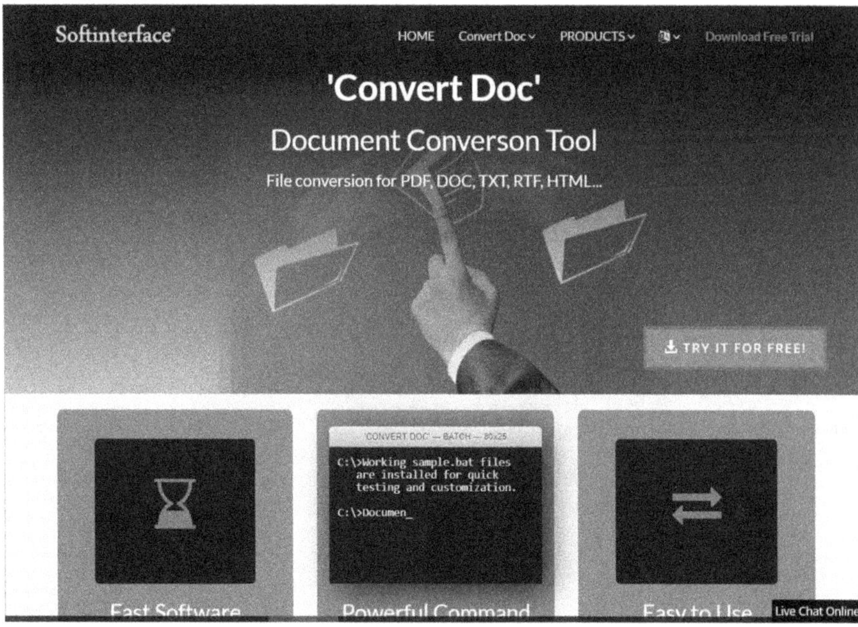

Fig. 6.6 The 'Convert Doc' website

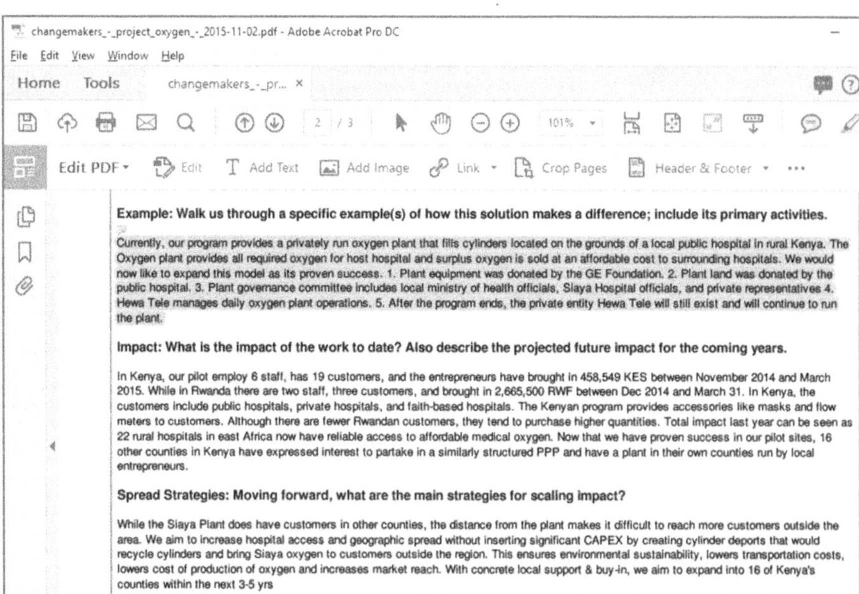

Fig. 6.7 A screenshot of PDF highlights

PDF-Xchange Viewer is required. After highlighting the selected text in RQDA, save the PDF file. In the following Chap. 7, we explain more about how to launch RQDA and how to import files for analysis into RQDA.

References

Cajaiba-Santana, G. (2014). Social innovation: Moving the field forward. A conceptual framework. *Technological Forecasting and Social Change, 82*, 42–51.

Mumford, M. D. (2002). Social innovation: Ten cases from Benjamin Franklin. *Creativity Research Journal, 14*(2), 253–266.

Nicholls, A., Simon, J., Gabriel, M., & Whelan, C. (Eds.). (2015). *New frontiers in social innovation research*. Berlin, Germany: Springer.

Phillips, W., Lee, H., Ghobadian, A., O'Regan, N., & James, P. (2015). Social innovation and social entrepreneurship: A systematic review. *Group & Organization Management, 40*(3), 428–461.

Terwiesch, C., & Ulrich, K. T. (2009). *Innovation tournaments: Creating and selecting exceptional opportunities*. Boston: Harvard Business Press.

Chapter 7
Starting RQDA

Abstract In Chap. 6, we discussed how to collect and prepare data files for qualitative analysis in RQDA. In this chapter, we demonstrate and discuss how to start and launch RQDA. This chapter will show you (1) how to start RQDA, (2) how to create a new project and manage your project(s) and (3) how to import files into RQDA.

Keywords Starting RQDA · Creating new project · Managing projects · Importing new files · RQDA package

7.1 Starting RQDA

Users who installed RQDA via R or RStudio platforms, take the following steps to launch RQDA:

Step 1: Start R or RStudio
Step 2: Go to the R Console or R Script in RStudio and type the syntax (see Fig. 7.1):

```
library(RQDA)
> RQDA()
```

Step 3: Run the Syntax in R Console by pressing Enter

The **library(package)** is a function in R that loads the names of packages available inside an R workspace. Here the syntax *library(RQDA)* means to load a package named RQDA, and after you press Enter you will see a message saying that the package(s) is being loaded (as shown in Fig. 7.2). If the package does not exist, an error message will appear. The RQDA package can only load if it has

Y. Chandra, L. Shang, *Qualitative Research Using R: A Systematic Approach*,
https://doi.org/10.1007/978-981-13-3170-1_7

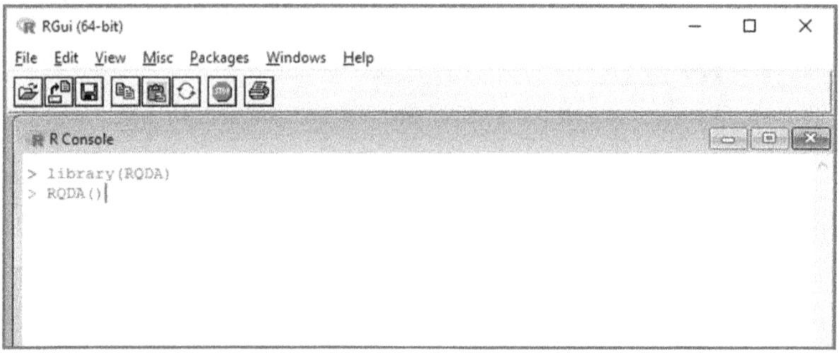

Fig. 7.1 Entering R syntax in R

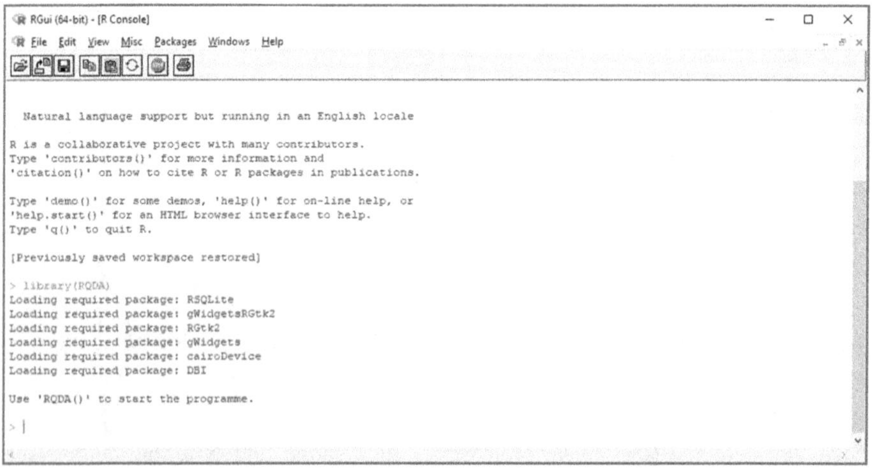

Fig. 7.2 Loading RQDA package in R

been installed in R. Therefore, if you try to launch RQDA, and fail, go to Chap. 5 and follow the steps in Sect. 5.1. to install RQDA properly in your R or RStudio platform.

If the RQDA package is successfully loaded on your R platform, a message will appear: *Use 'RQDA()' to start the programme*. After you type the syntax *RQDA()*, a RQDA panel should then open in a separate window, ready for data analysis (see Fig. 7.3). (Note that you may also use the same steps described above to launch other R packages by simply replacing '*RQDA*' with the name of your target package.)

When RQDA's main panel is opened, nine menus appear on the right side of the panel (starting with **Project** and moving down to **Settings**), each of which has a different function. We explain the most important menus as we go forward.

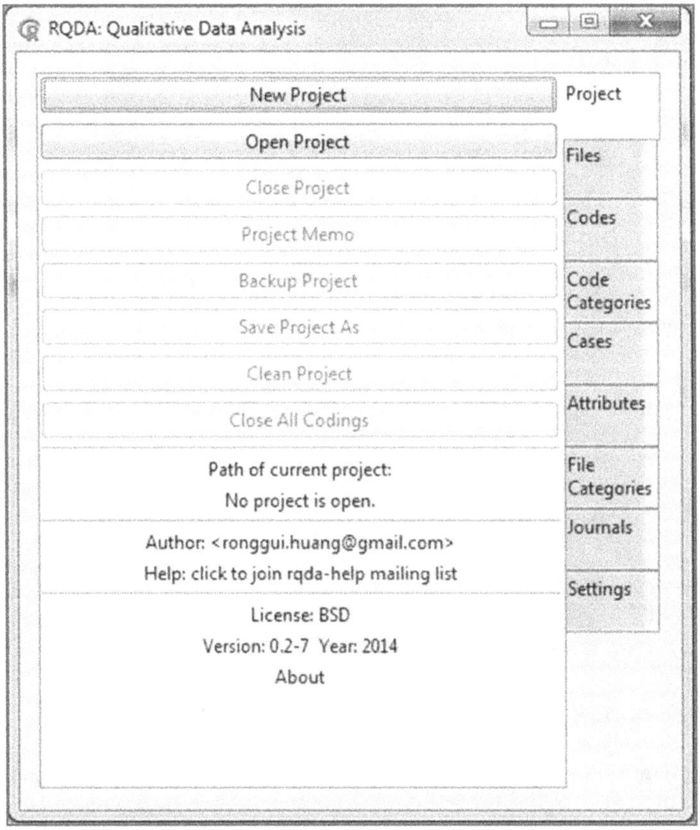

Fig. 7.3 RQDA's main panel with various menus and functions

Windows users without R or RStudio installed can directly download a .bat file (i.e., a batch file, filename extension .bat, is used in DOS and Windows) to start RQDA as demonstrated in Sect. 5.2. First, extract the *rar* file to, say, c:\ (or other location that you prefer). Then you can launch RQDA by double-clicking the batch file 'RQDA' in the extracted folder. An R console will open and a RQDA panel should pop up in a separate window within seconds. If it does not open automatically, please try again. If it still fails, you can use an alternative manual method to start it. To do so, go to the **R Console** that opens after you double click the batch file *RQDA.bat*, and choose *Packages* from the menu. A sub-menu should appear; click *Load package…* as shown in Fig. 7.4.

After you click on *Load package…*, a menu with names of different packages will pop up in a new window; select *RQDA* (see Fig. 7.5). Then press *OK*, and a RQDA panel should open within a few seconds.

Alternatively, if the RQDA panel does not pop up automatically after double clicking 'RQDA.bat', you can directly type syntax *RQDA()* in the R Console to

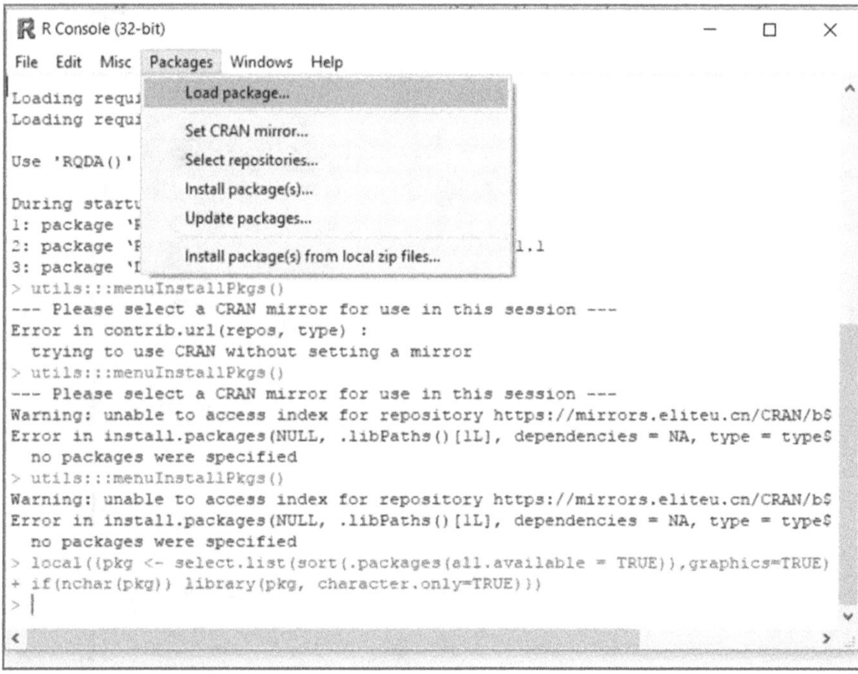

Fig. 7.4 Starting RQDA in RQDA batch file

launch RQDA. Note that the RQDA package is already pre-loaded in the Console and therefore you do not need to type *library(RQDA)* to load the package.

7.2 Creating a New Project in RQDA

To start a new qualitative research project, select *New Project* from the **Project** menu, type a new, specific project name and select the location in which to save your RQDA project, and then click *OK*.

For example, we created a new project in RQDA for our illustrative project and named it *SI Strategies + Impacts in Healthcare* and saved this project in the Desktop (see Fig. 7.6).

After you click OK, a new project is created as a .rqda file in the location you chose to save the file. Later when you want to open this project, you can simply click on *Open Project* in the **Project** menu in RQDA, find the .rqda file and click *OK*; the file will be loaded into RQDA (see Fig. 7.7). When the project is open, you can find its path information in the **Project** menu (see Fig. 7.8), below the *Close All Codings* option. This information shows the project that is currently open and where it is stored.

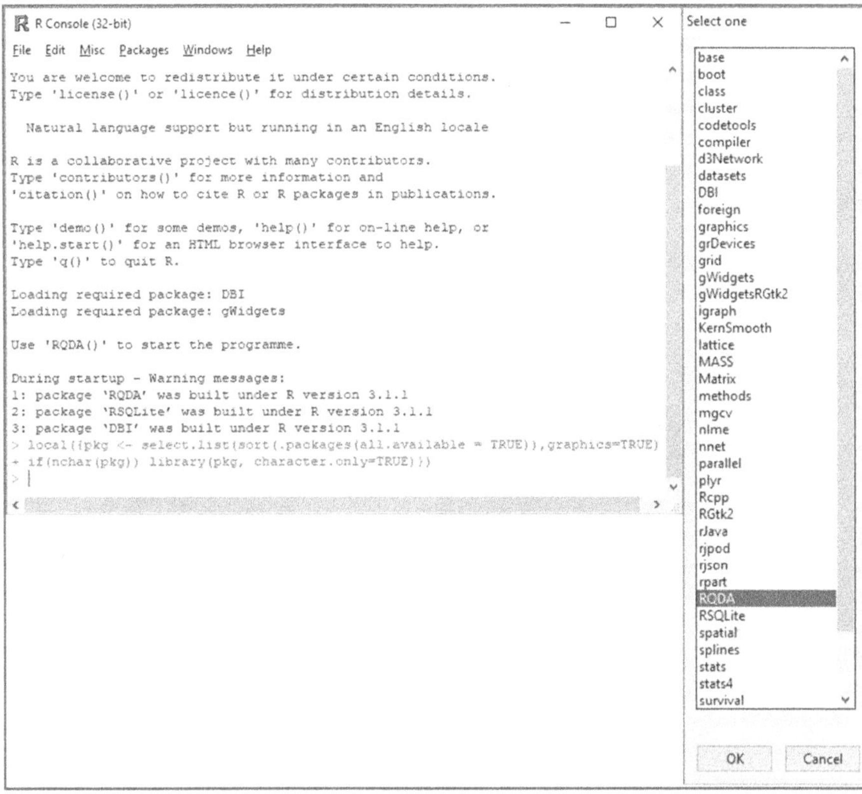

Fig. 7.5 Loading RQDA package in R console

To close the project, click *Close Project* and your project will be closed. You may create as many projects as you want by selecting *New Project* (e.g., you may want to work on another project that studies the social networks of social innovators). However, you can only work on one project at a time in RQDA.

Also, if you want to back up your project, click on *Backup Project* and a new backup file will appear in your designated storage location as shown in Fig. 7.9.

As you can see from Fig. 7.9, the first .rqda file is the backup project that we just created, and the second file is the original. The number extension (17453008012018) after the name of the first .rqda file indicates the exact date and time that this backup was created. By using the backup function in RQDA, you can easily track your progress and the changes you have made.

The **Project** menu also has a *Save Project As* option, which enables users to save the project under a new name in a chosen location (similar to the *Save as* option in Microsoft Word). RQDA's *Backup Project* function enables a backup but not under a different name.

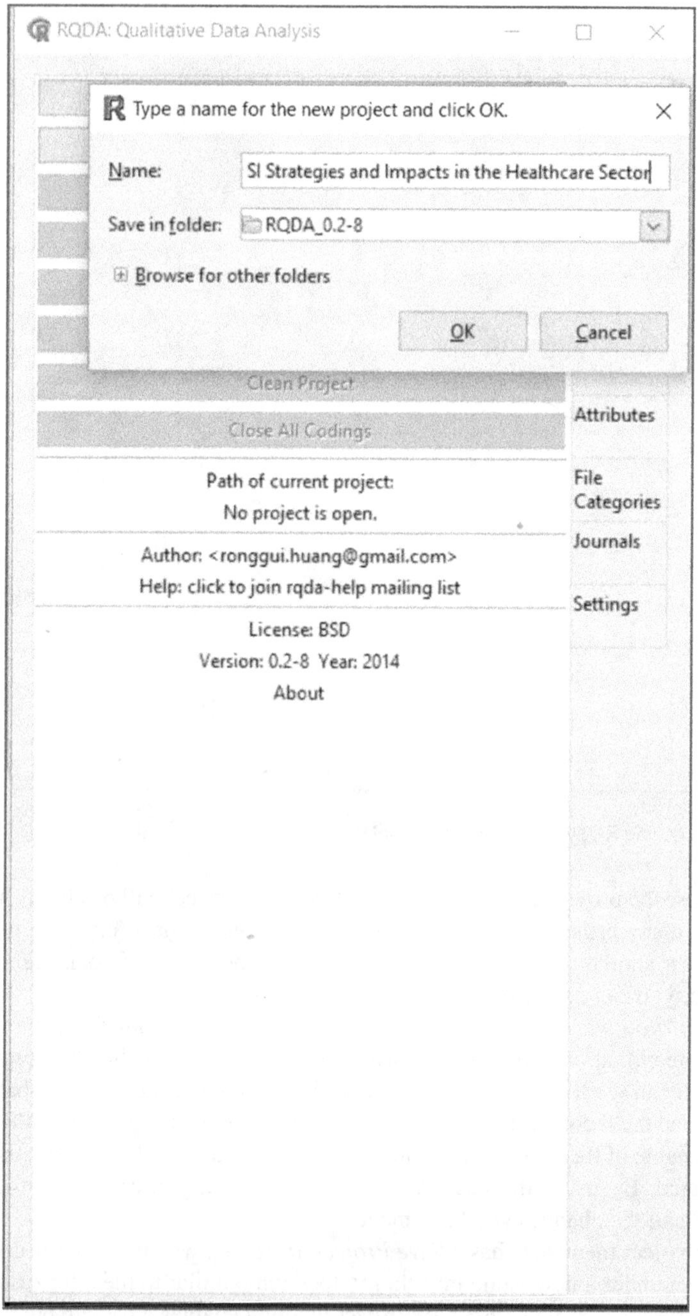

Fig. 7.6 Creating a new project in RQDA

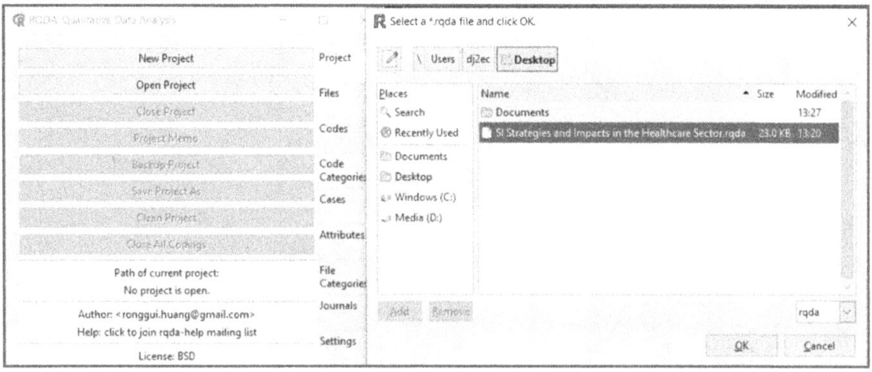

Fig. 7.7 A screenshot of opened saved project in RQDA

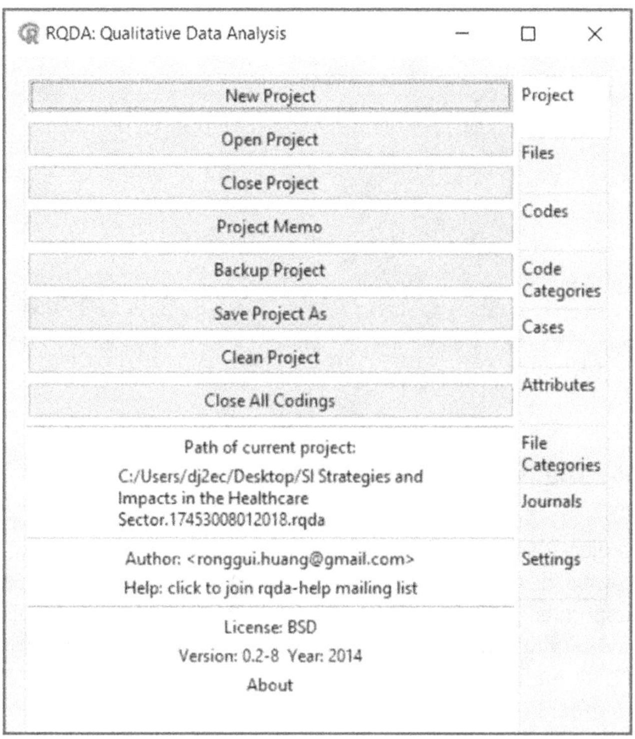

Fig. 7.8 Path information of an open project

Fig. 7.9 Backup project

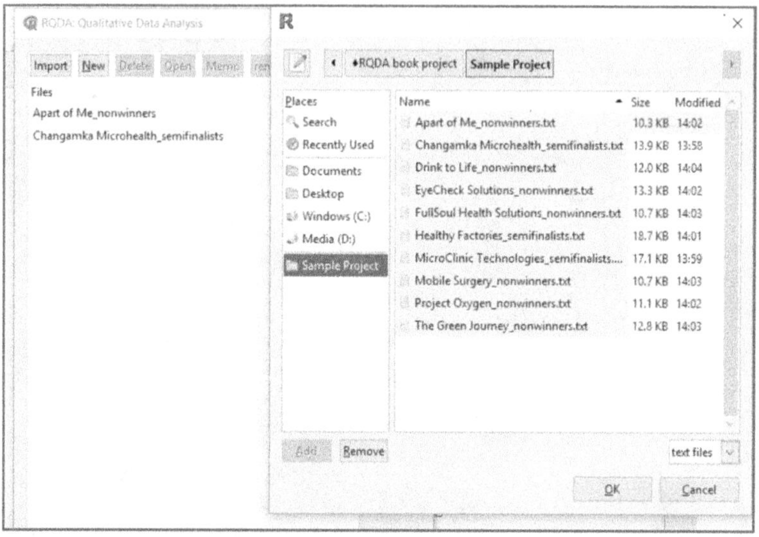

Fig. 7.10 Importing new data files into RQDA

7.3 Importing New Data Files into RQDA

Once a new project is successfully created, users must import new data files, which must be plain text or *.txt* files or *highlights* in PDF files). After this, users can start the coding process. To import the plain data .txt files for coding, select the *Import* button from the **Files** menu and select the file that will be imported into RQDA (as shown in Fig. 7.10).

Each .txt data file imported will be saved as a new file in RQDA. RQDA only allows you to import one file each time.

One easy way to locate the folder where you saved all imported .txt files is to use a **bookmarks** function, by selecting the target folder and clicking *Add* in the **Import** window. For example, we saved all our .txt files in the **Sample Project** folder and we added this folder to **Bookmarks**. The next time we want to import new files, we can quickly locate the storage folder by clicking *Sample Project* in the bookmark (under **Places**; see Fig. 7.11). If you no longer use this folder to store new files, you can delete it by selecting the folder under **Places** and clicking *Remove*.

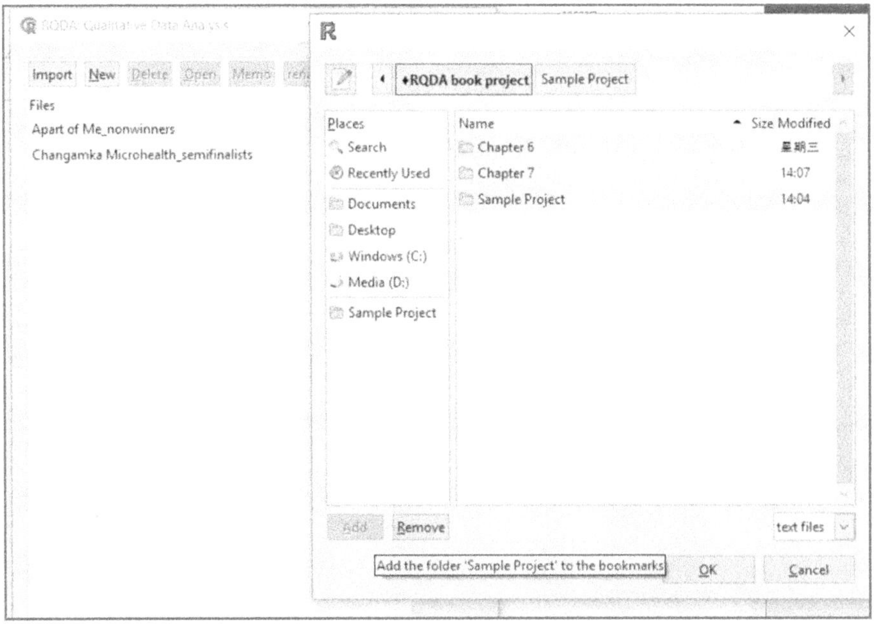

Fig. 7.11 Adding folders to bookmarks

Alternatively, users can add new files by right-clicking and selecting *Add New File* in the **Files** tab and a new window will pop up that enables users to copy and paste the text content directly to the text entry box in the window. In this way, you save time converting document formats (as shown in Fig. 7.12). Be sure to save the file by clicking on *Save and Close* button after you finish copying it to the text entry box to avoid potential data loss.

So far, we have discussed two ways of importing data into RQDA. These methods are easy to use when you have just a few files, since RQDA only allows users to *import one file at a time*, but it can be quite time-consuming to import a large number of files. To work efficiently, users can import a batch of files simultaneously using R syntax. For example, to import two individual plain text (.txt) files, follow these steps:

1. Open or create a new project file (go to Sect. 7.2. for details).
2. Go to the R console and type the syntax:

```
Files <- list("Text1"="Text1 content.","Text2"="Text2 content.")
write.FileList(Files)
```

3. Press *Enter*.

The "Text1" and "Text2" referenced in the above syntax are the file names of the user's target data files, and "Text1 content" and "Text2 content" refer to the content

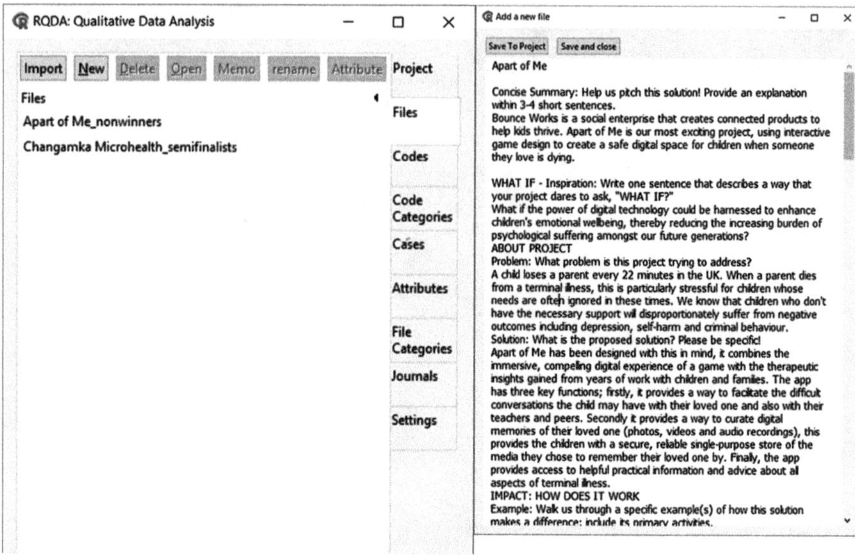

Fig. 7.12 Showing the content of a qualitative data file

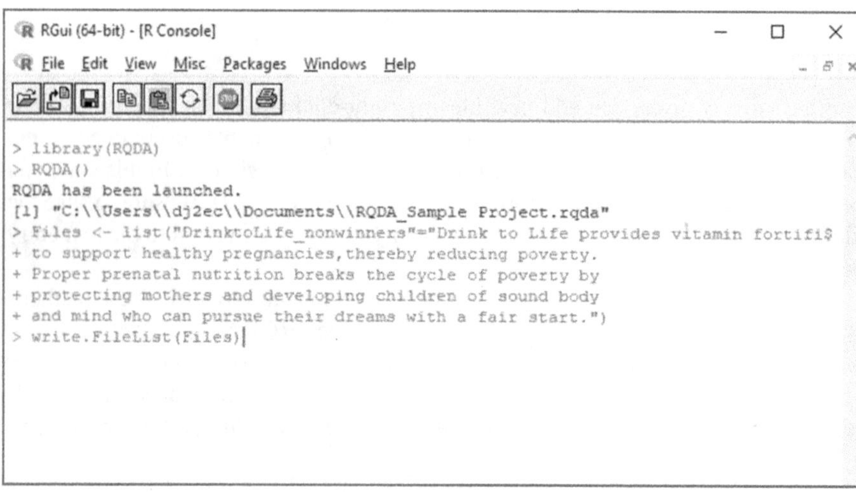

Fig. 7.13 Importing a batch of textual data files into RQDA

of the files that the user wants to import and analyze later. Next, after pressing the *Enter* button, two new data files, "Text1" and "Text2", should appear in the RQDA panel. The new project created is saved *automatically* by RQDA. For instance, in our sample project, we want to import two textual data files for the SI project 'Drink to Life' and 'Mobile Surgery' into RQDA. To import them together, type the above syntax and use the actual file names and text contents to replace 'Text1', 'Text2', 'Text1 content' and 'Text2 content' (as shown in Fig. 7.13).

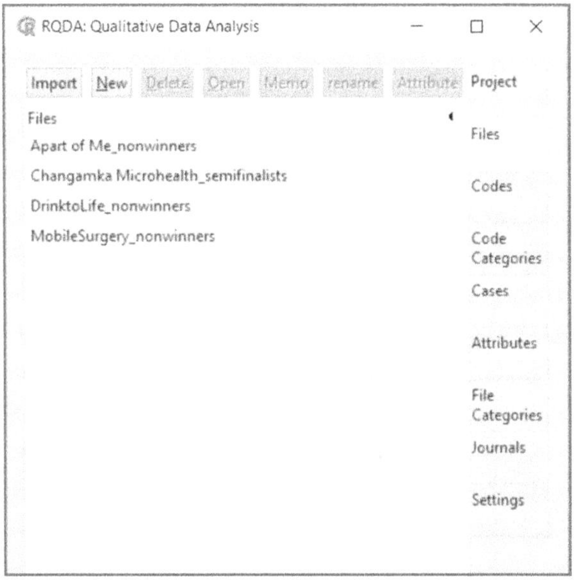

Fig. 7.14 Newly created textual data files in RQDA

After we press Enter, we can then find two newly created data files in the RQDA Files menu ('DrinktoLife_nonwinners and MobileSurgery_nonwinners; see Fig. 7.14). Please note that if you have questions about the R function *write. FileList()*, you can refer to the RQDA Help page by typing syntax *'?write.FileList'* in the R console (this also applies to other R functions).

Lastly, as we mentioned in Chap. 6, RQDA also enables users to import color highlights in PDF files. Here we use the SI profile *Project Oxygen* as an illustration. The SI project provided supplementary information in the PDF format in the Changemaker website. First, we downloaded the PDF file from the website (https://www.changemakers.com/makingmorehealth/entries/project-oxygen). Note that you must use an appropriate PDF reader to open your PDF files, such as Adobe Acrobat Pro, Foxit Reader, PDF Xchange Editor, SlimPDF Reader or PDF Xchange Viewer. In our sample project, we used Adobe Acrobat Pro DC for illustration. After you open the PDF file, you must then set your commenting preferences in the **Preferences** menu (Click on **Edit** tab to find **Preferences**; please note that the location of the **Preferences** menu may be different in different PDF readers) In the **Preferences** menu, find **Commenting** under **Categories**, and select *Copy selected text into Highlight, Strikethrough, and Underline comment pop-ups* by simply checking the box. By doing this, the highlights you make in the PDF file will be recognized in RQDA (see Fig. 7.15). If you use another PDF reader, also check this option in the **Preferences** menu.

Next, we used the *Highlight text* tool in the PDF reader to highlight important text that we want to import into RQDA for analysis (see Fig. 7.16). If your preferences settings are correct, you will see a note containing the highlighted text pop up when you place the cursor over the highlighted text. You should save all changes in your PDF files before you import data into RQDA.

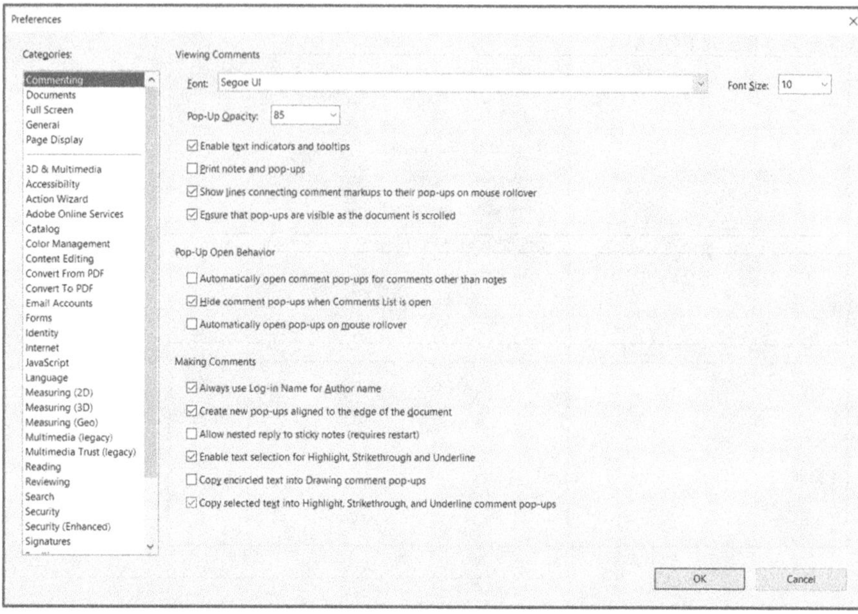

Fig. 7.15 Setting the commenting menu in Adobe Acrobat Pro

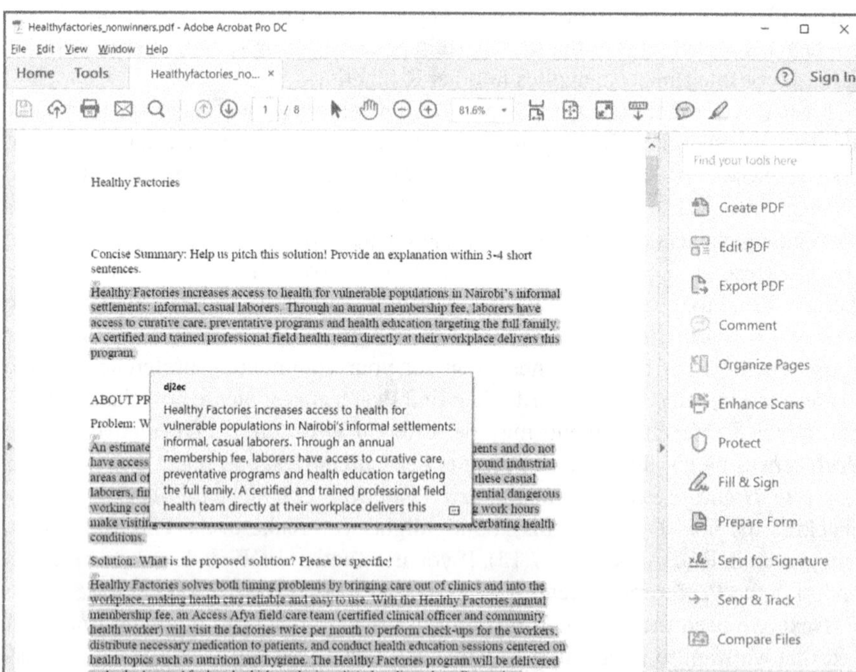

Fig. 7.16 Highlighting parts of a text in PDF

To import the color highlights in PDF files into RQDA, follow these steps:

1. Open an R Console
2. Install R package 'rJava' by typing syntax:

```
install.packages("rJava")
```

3. Load 'rJava' in R by typing syntax:

```
library(rJava)
```

4. Upgrade RQDA if your installed version is <0.2-5 by typing syntax:

```
install.packages("RQDA", repos="http://r-forge.r-project.org")
```

After you finish these steps, you can launch RQDA and open a RQDA project. Go to the **File** menu, right-click on the blank space and a new menu will pop up (see Fig. 7.17).

Fig. 7.17 A new pop-up menu in files tab in RQDA

Please note that the R package *rjpod*, which supports PDF highlights import in RQDA is no longer available in the latest version of R. If you are a non-R user, and you downloaded RQDA from the developer's (Huang Ronggui) shared cloud drive, you still can use this function as RQDA.bat file shared by Huang is based on an older version of R. To illustrate this function, we directly downloaded an RQDA 0.2-5 batch file from Huang's shared **OneDrive** folder and run it in the Windows operation system: https://onedrive.live.com/?id=41C1EE655922D297%21278&cid=41C1EE655922D297. This version of RQDA is running on R version 2.15.2, and it still supports users who wish to import PDF highlights to files. If you see an error like the following after you run the syntax: '*library(rJava)*', it is most likely that you did not install Java in your computer or you did not set JAVA_HOME (which means setting up the environment for Java) in R. The error information is as below:

```
        Loading required package: rJava
        library(rJava)
        Error: .onLoad failed in loadNamespace() for 'rJava',
details:
        call: fun(libname, pkgname)
        error: JAVA_HOME cannot be determined from the Registry
```

To fix this problem, you could follow the following steps:

- For those who do not have Java, you should download it and install it in advance before you load 'rJava' in R (for download, go to: http://www.oracle.com/technetwork/java/javase/downloads/jre8-downloads-2133155.html). Select Java Runtime Environment and after you successfully install Java, you need to set JAVA_Home by typing the following syntax in your R Console:

```
       Sys.setenv(JAVA_HOME='C:\\Program Files (x86)\\Java\\jre7')
        #you should change the library destination to the location
where your Java is installed#
```

Then try running the syntax '*library(rJava)*' again.

Fig. 7.18 Imported PDF highlights in RQDA

To import PDF highlights, you should right-click on the blank space in the **Files** tab and select **Import PDF Highlights via rjpod (selector)**, then select a PDF file with highlights. For illustration, we selected the PDF file that we created for an SI project Healthy Factories (see Fig. 7.16). After selecting the PDF file, a new file is created in the RQDA file menu. All highlights in the file will appear by double-clicking the new file and opening it as shown in Fig. 7.18. This imported file will have a file extension *.pdf*, while other files imported from plain text will have the .txt file extension. These file extensions will help you better manage your files as you can distinguish which files are imported from PDFs, text files or directly created in RQDA (for files that are directly created in RQDA, they do not have any extensions).

Chapter 8
Inductive Coding

Abstract The most important function of any CAQDAS platform including RQDA is the *coding* function. Coding assigns a meaning to a small body of text (e.g., a specific word or lexical item, a sentence, a phrase or paragraph) using a label (usually one to a few words, such as *Used_digital_technology* or *Formed_a_peer_support_group*) that best represents the text. RQDA allows *inductive* and *deductive* coding, and its operation is intuitive. Our focus here is consistent with the Gioia approach to inductive coding. In RQDA, coding is quite simple and intuitive. This chapter discusses the inductive coding process in qualitative research and how to do so in RQDA. First, we briefly introduce the concepts of inductive and deductive coding and how they differ. Then we provide detailed instructions on the inductive coding process in RQDA by analyzing the data from our sample project, *SI Strategies and Impacts in the Healthcare Sector*, as an illustration.

Keywords Inductive coding · Deductive coding · Code labels · Codes · First-order codes

8.1 Inductive and Deductive Coding

Inductive coding refers to a data analysis process whereby the researcher reads and interprets raw textual data to develop concepts, themes or a process model through interpretations based on data (Thomas 2006; Boyatzis 1998; Corbin and Strauss 1990). As its name suggests, the inductive coding approach generally relies on *inductive reasoning*, in which important themes, topics or models emerge from the raw data through repeated examination and comparison. It begins with organizing the raw data, which might come from news articles, interview transcripts, marketing reports, academic articles or other material through a process known as *first-order coding* or *open coding*. Through first-order coding, a researcher closely reviews the data, makes notes, combines the data into broader themes and theoretical dimensions. Inductive coding requires the marking of passages and sentences of a text with a code.

Y. Chandra, L. Shang, *Qualitative Research Using R: A Systematic Approach*,
https://doi.org/10.1007/978-981-13-3170-1_8

It is a data-driven approach in which '*the research begins with an area of study and allows the theory to emerge from the data*' (Strauss and Corbin 1998, p.12). Researchers using an interpretivist-constructivist approach prefer inductive coding (Corbin and Strauss 1990; Gioia et al. 2013) because it enables users to code a textual unit (e.g., words, sentences or paragraphs) using labels that are close to the data without influence from existing concepts, constructs, or theories. For example, using our illustrative SI data, the sentence *"Changamka Microhealth seeks to drastically lower the cost of health insurance by developing a product based partly on phone contact"* (a statement of the organization's strategy) can be coded as *Alternative, Cost Cutting Health Care Approach.*

For most researchers, the primary reason for using the inductive coding approach is to analyze data in areas with *limited knowledge* (e.g., an emerging field of research, a new aspect or dimension of interest within an existing field, or a knowledge gap in a concept, approach, or phenomenon). This type of data analysis does not require any predetermined concepts or theory about topics or codes.

Deductive coding refers to a coding process aimed to test whether data are consistent with prior assumptions, theories, or hypotheses (Thomas 2006; Scheunemann et al. 2015; Fereday and Muir-Cochrane 2006; Bazeley and Jackson 2013). For example, if a researcher has a preconceived notion that innovativeness (Verhees and Meulenberg 2004) will influence product innovation processes in a positive way, then the researcher may code SI strategies that contain new product development and use cutting-edge technologies *as evidence of the 'innovativeness' construct.*

For example, Apart of Me, a grassroots organization in England that seeks to address mental health crisis among youths using digital technologies and games, positions itself as a social innovator that provides "a safe virtual space that helps families find their way through grief … combines the immersive, compelling digital experience of a game with the therapeutic insights gained from years of work with children and families by using interactive game design to create a safe digital space for children when someone they love is dying" (https://bounce.works/apartofme/). This statement can be coded deductively as a data that shows the presence of *innovativeness* construct.

The main difference between the inductive and deductive coding process is that a deductive approach is aimed at testing theory and usually begins with a hypothesis or proposition (Fereday and Muir-Cochrane 2006). A deductive coding process involves a top-down coding approach. Usually, a coding scheme or a coding manual (with precise measurement scales such as 'yes' or 'no', or metric information such as 'sales in numerical value') are preconceived based on existing theories, themes or models to guide the deductive coding process. Once the coding scheme (a complete list of all constructs and how they are measured) is ready, the researcher will apply the codes to the text. Most qualitative research refers to an inductive coding process rather than a deductive one.

In our sample project, we adopted Gioia approach's inductive coding approach and coded the organizational material of participants of *"Making More Health: Co-Creating a Healthier World"* (or MMH), an Ashoka Changemakers' SI competition in healthcare, to identify the strategies and impacts of SI projects in the healthcare sector. To demonstrate the Gioia-based qualitative research process using RQDA, we first imported 10 organizational texts consisting of three semi-finalists and seven non-winners (from the MMH competition) to RQDA, along with their supplementary data. Figure 8.1 shows the stages of the inductive coding process.

In our sample project, we used RQDA to complete the first five steps as shown in Fig. 8.1 and used Microsoft Visio to draw the data structure for optimal visual-

Fig. 8.1 Steps of data coding

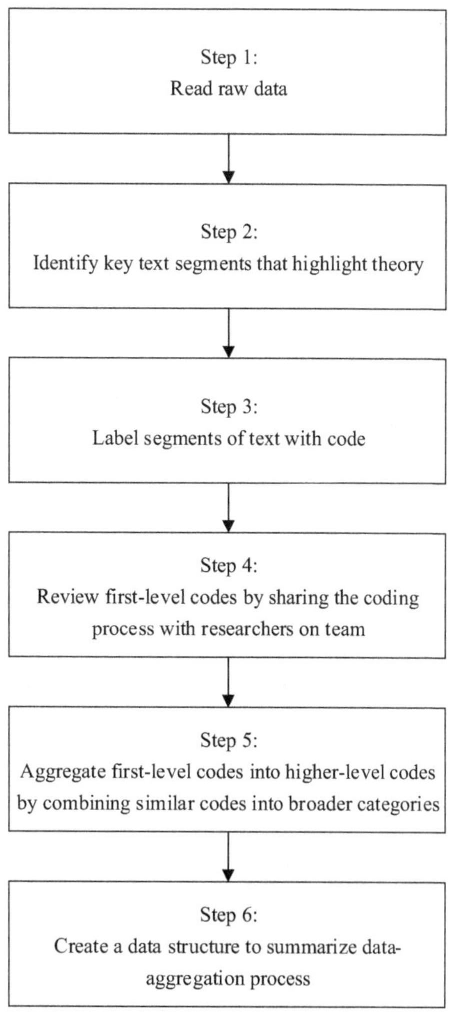

ization. In the following section, we discuss the detailed steps of how to assign codes to text segments in RQDA and other useful functions for qualitative data analysis.

8.2 Coding in RQDA

As mentioned above, coding is one of the most important functions of RQDA. The coding process is straightforward. After you import files (e.g., interview transcripts, texts of news stories), simply double-click on a file that you want to analyze under the **Files** tab (see the Files function in Fig. 7.8) and then a new window will appear showing the content of the selected file (see Fig. 8.2).

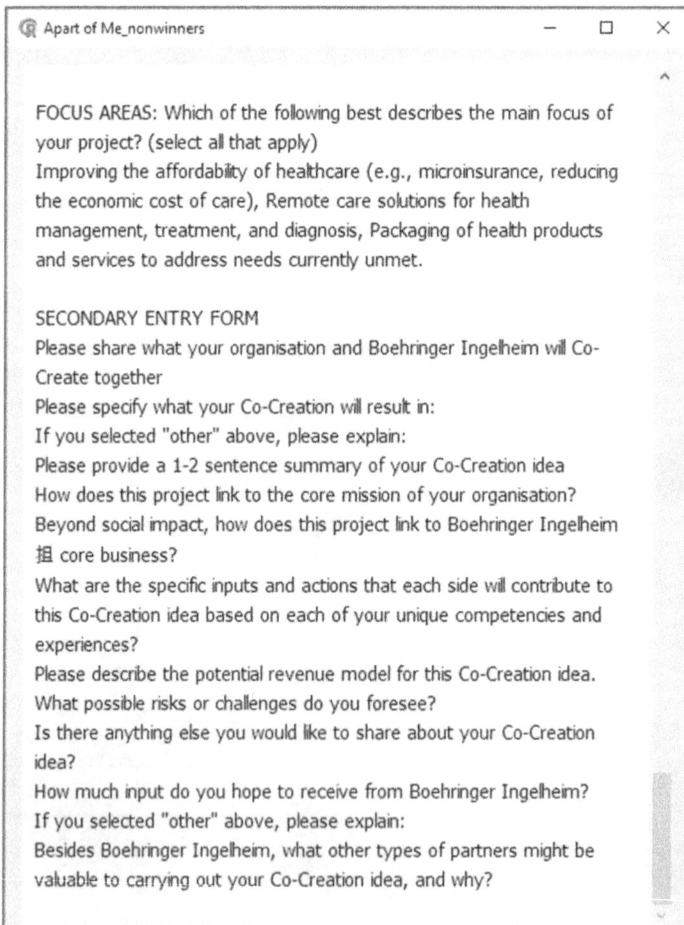

Fig. 8.2 A file window in RQDA

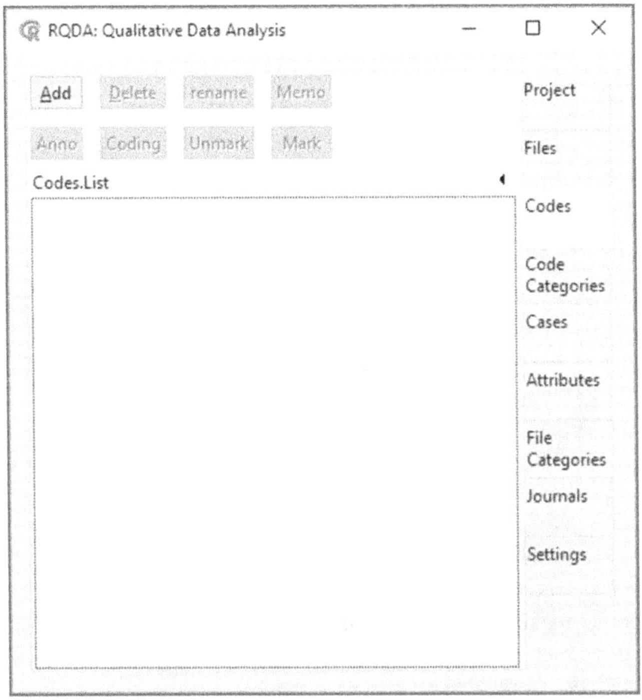

Fig. 8.3 A codes tab in RQDA

Under the **Codes** tab are eight buttons (*Add, Delete, Rename, Memo, Anno, Coding, Unmark* and *Mark*) that allow you to add, edit, or delete your codes (see Fig. 8.3).

To code the selected file, we assigned codes to text segments related to the strategies used by social innovators and the impacts of their SI projects following these steps:

1. Carefully read the text and find useful or meaningful words or sentences that can be coded as first-order concepts (see Sect. 2.2. for the definition of first-order concepts).
2. Click the *add* button in the **Codes** tab to input a new code.
3. Select sentences or paragraphs to be coded.
4. Click the *Mark* button on the **Codes** tab and apply it to the words or sentences to be coded.

There are no hard and fast rules on how much to code in making first-order codes. The usual practice is to code at a sentence level. Researchers also code at the paragraph level. There is nothing wrong to code at a phrase or keyword level if a researcher thinks that it contains significant meaning or revelation or discovery of a concept. In our research, we coded at *all levels*: sentence, paragraph, and key phrase or key word.

Another frequently asked question is how does one know if the codes that he/she created are detailed enough, or how broad should the codes be? Our advice is for researchers to use *as detailed as possible* code labels. This tactic will prevent *information loss*, which often happens when a researcher uses too broad code labels at the first-order code, leaving little information to be aggregated to create second-order codes.

Creating *multiple code labels* that represent multiple yet nuanced meaning is also a good practice. For instance, the sentence "We [a young couple] don't want to have a child…but we love pets…we treat Pluto [a baby Maltese] like a child" can be coded as ***"Pets_as_substitute_for_child"***, ***"Pets_as_Child"***, and ***"Young_Couples_Prefer_Pets_than_Kids"***.

Another important advice in coding is to code *conservatively* by including code labels that express what was explicit in the data, and not inferring *intentionality*. The examples above show code labels that are coded conservatively and did not include any intention of the text producer. A research could add the intention of the text producer in a code, but it may mislead readers and bias the findings. Using the examples of the married young couples who prefer pets than children above, a researcher could make his/her own inference that expresses the intention of the couple not to have children, such as **"High_Living_Costs"**, **"Homes_Are_Unaffordable"**, or **"Our_Work_Life_Is_Too_Stressful"**. A conservative approach is always a better strategy in conductive any type of research.

For example, in our illustrative project, we know that the SI project "Drink to Life" seeks to solve malnutrition problems in developing countries by providing vitamin fortified beverages to pregnant and lactating women support healthy pregnancies. We created a first-order concept called "***Worked_to_support_healthy_pregnancies***" for text that read: "*First, rather than cure, we will prevent malnutrition. The majority of the above economic and social costs stem from a failure to nourish infants during the critical period between conception and their second birthday, during which damage from malnutrition is irreversible. Interventions such as PlumpyNut are useful but often administered too late in a child's lifecycle. Drink to Life's prenatal beverages support healthy pregnancies and will nourish a new generation during critical development*". As shown in this example, we coded the entire **paragraph** and identified the activities of Drink to Life.

We can also code the text above based at the **sentence** level. For example, "*First, rather than cure, we will prevent malnutrition*", can be coded as "***Prevent_Malnutrition***". Likewise, the sentence "*Drink to Life's prenatal beverages support healthy pregnancies and will nourish a new generation during critical development*" can be coded as ***"Prenatal_Beverages_For_Healthy_Pregnancies"*** (which is the strategy or intervention to the infant malnutrition problem).

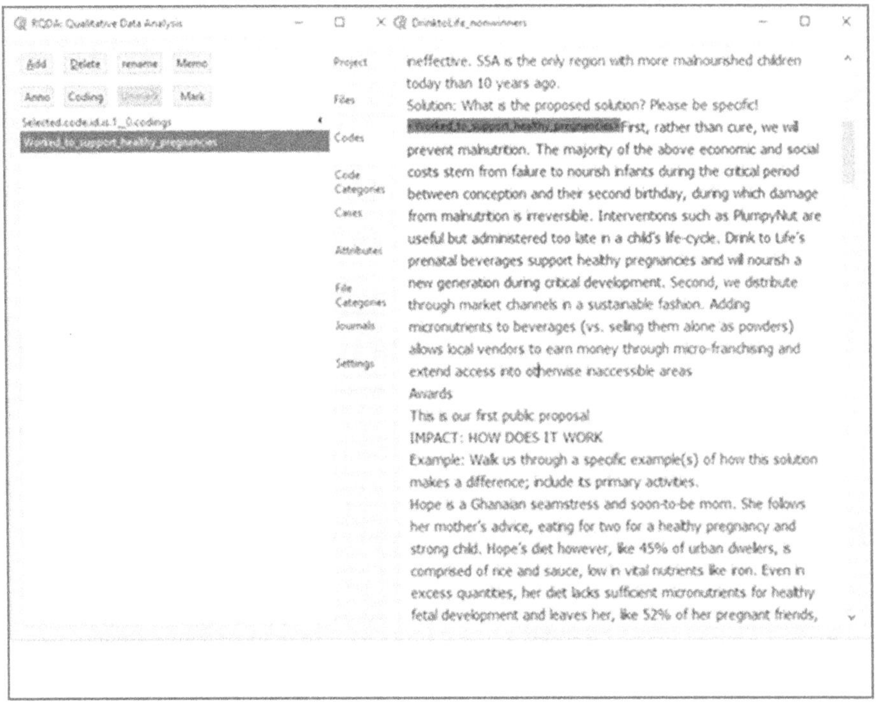

Fig. 8.4 Sample coded text in RQDA

We can also capture multiple meanings of a sentence using multiple code labels. For instance, the sentence *"The majority of the above economic and social costs stem from a failure to nourish infants during the critical period between conception and their second birthday, during which damage from malnutrition is irreversible. Interventions such as PlumpyNut are useful but often administered too late in a child's lifecycle"*, can be coded as: ***"Failure_to_Nourish_Infants_in_Critical_Period", "Malnutrition_is_Irreversible", "Existing_Interventions_Are_Given_Too_Late".***

After you add a new code, you may find it in the **Codes** tab under the *Codes.List*. The coded text segment is automatically highlighted in color and a coding label appears before the coded text as shown in Fig. 8.4.

As shown in Fig. 8.4, the coded sentences are highlighted in blue and the selected texts are coded. The code label appears in front of each coded texts. The color of the first-order coded concept can be changed to suit your preferences. You may select your preferred color under RQDA's **Color for Coding** menu in the **Settings** tab. RQDA offers hundreds of colors (see Fig. 8.5).

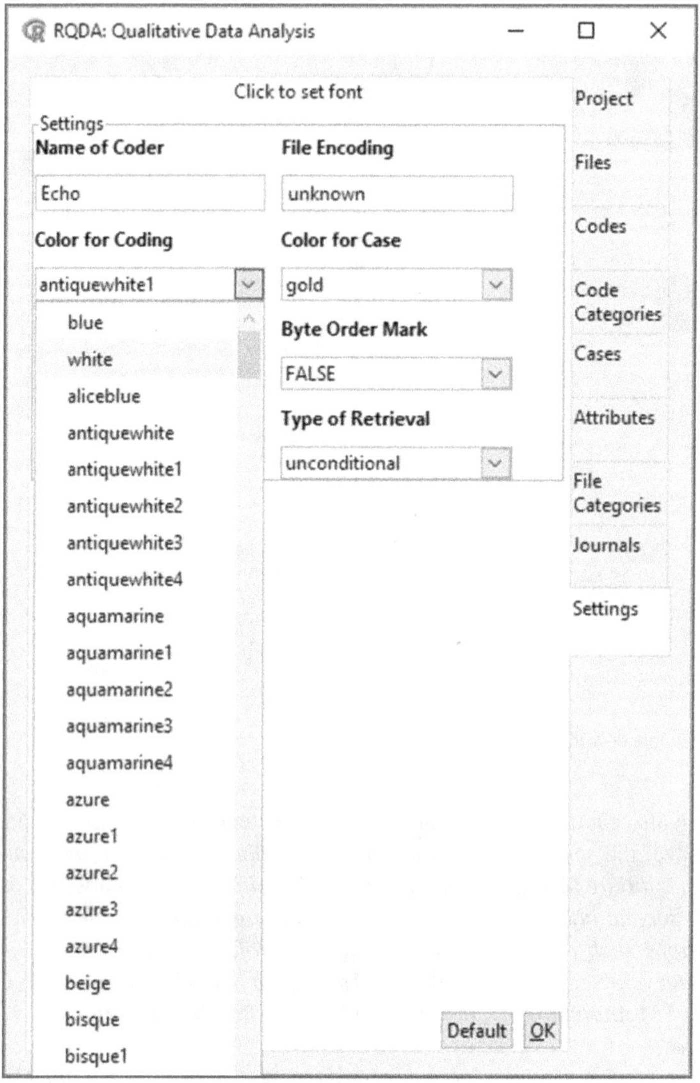

Fig. 8.5 Changing color for coding in RQDA

As you interpret and code all the text files imported to RQDA for a project, you can either create new codes when new concepts emerge or continue using existing codes to code other texts that fall under the existing code labels. You can stop creating code labels when no new concepts emerge.

Researchers working within the interpretivist-constructivist school can code a phrase or sentence or paragraph into more than one first-order concepts if the phrase or sentence or paragraph can plausibly mean more than one thing (e.g., "giving cash

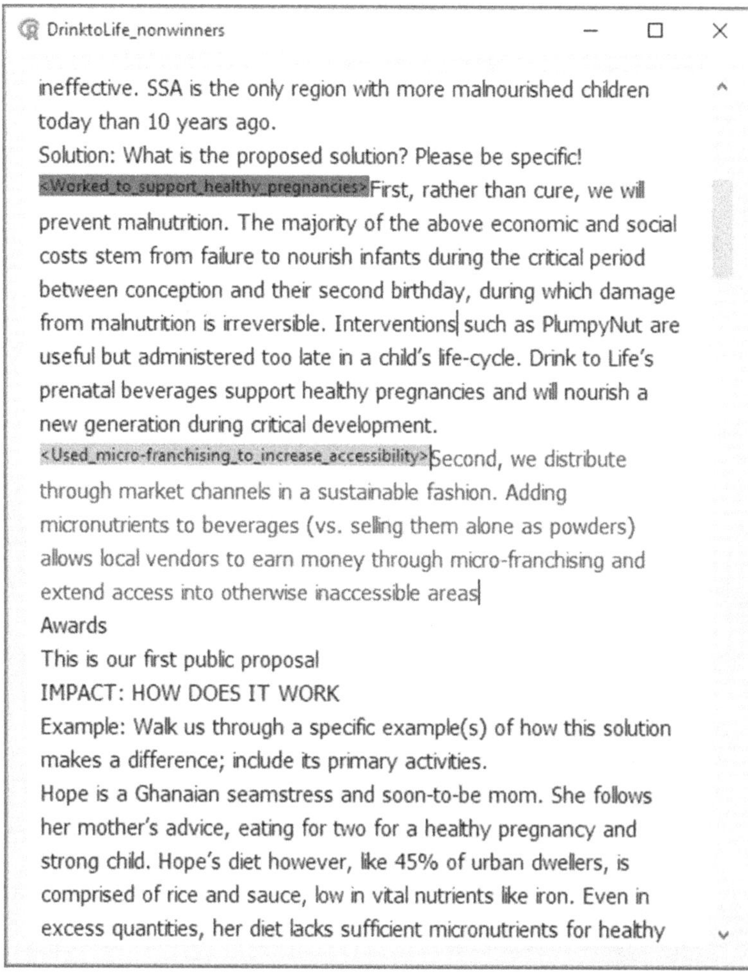

Fig. 8.6 Highlighted codes in RQDA

to a beggar doesn't not solve the problem" can be coded as "*avoid_dependency*" as well as "*giving_cash_problematic*"). There is no limitation on the number of codes that can be created in RQDA, and different codes can be shown in different highlight colors in the **Files** window as shown in Fig. 8.6. In this way, researchers can easily gain an overall understanding of how different codes are applied throughout the text.

RQDA automatically highlights text segments coded under the same code in the same color to ease understanding. Notably, users can delete or rename any code in

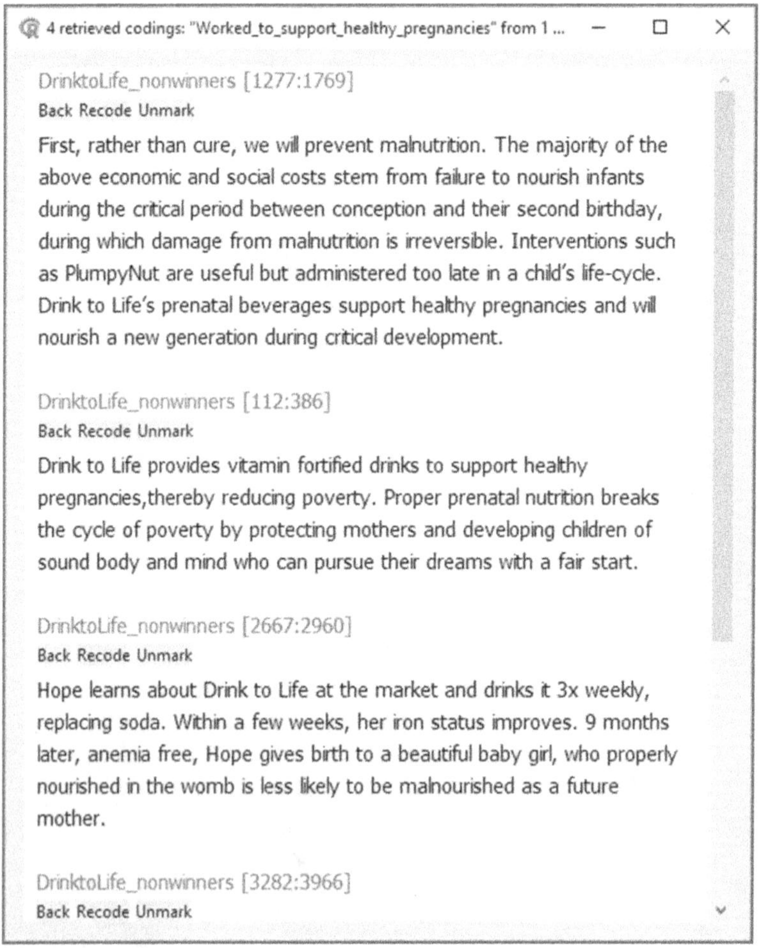

Fig. 8.7 Viewing coded text segments of a code in RQDA

the **Codes** menu. To delete or rename a code, select the code and click on the **Delete** or **Rename** button, respectively. In addition, users can *undo* coding by selecting the coded text and clicking the **Unmark** button in the **Codes** menu, which will remove the coding label and its color highlights.

Moreover, RQDA enables users to review all coded text segments under a particular code at any time by double-clicking the code from the **Codes** menu. From here, a new window appears with all the coded text segments under the selected code. For example, to view all coded text segments under the code "Worked_to_Support_Healthy_Pregnancies," we double-clicked on the code in the **Codes** menu and a new window popped up as shown in Fig. 8.7.

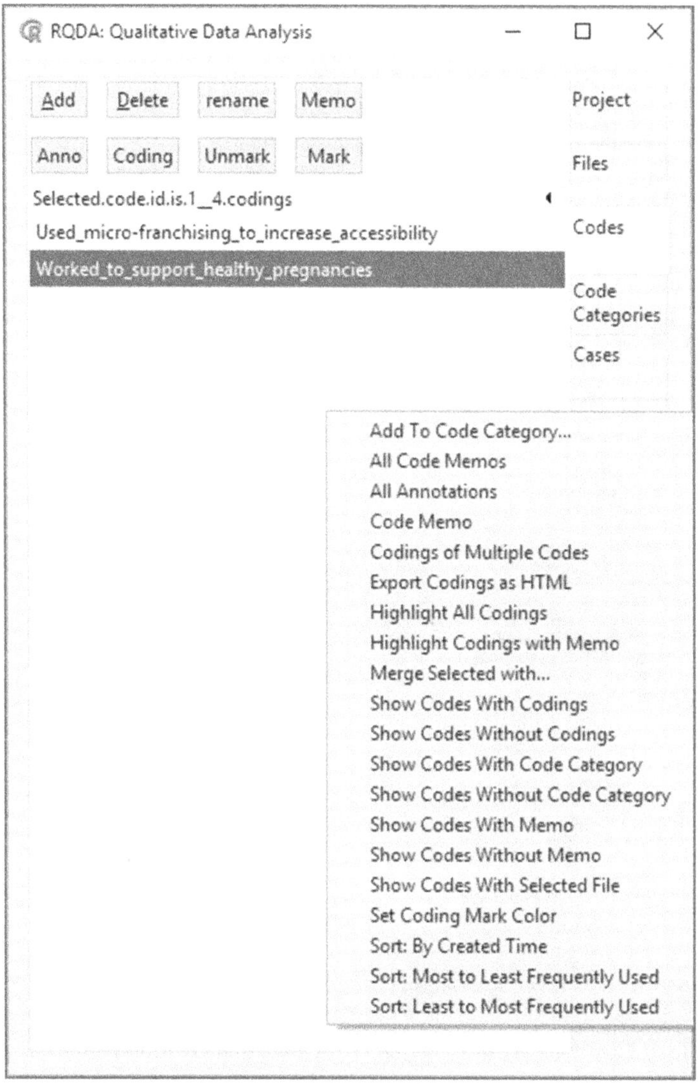

Fig. 8.8 Setting conditional retrieval in RQDA

The screenshot in Fig. 8.7 shows all the codes from different files under the code "*Worked_to_support_healthy_pregnancies.*" Users can open an individual file by clicking the *Back* button below the file name (e.g., Apart from Me_non-winners). *Conditional retrieval* is also permitted. For example, users can either show codes with or without the coding category or sort codes based on the time at which the codes were created or their frequency by right-clicking on the Codes menu (see Fig. 8.8).

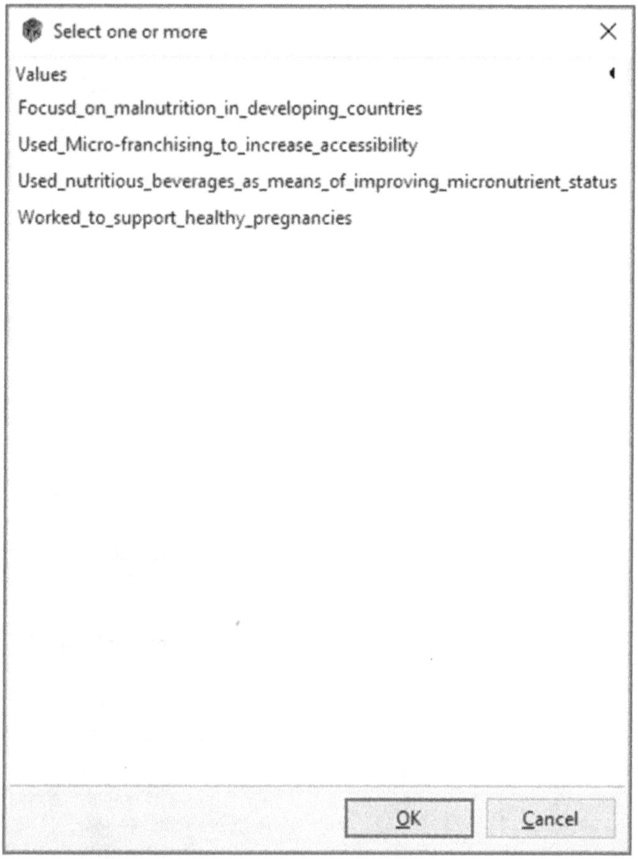

Fig. 8.9 Merging codes in RQDA

Moreover, you can recode or unmark the coding by simply clicking the *Recode* or *Unmark* button in the **Codes** menu. Alternatively, you can unmark the coding within the **Files** menu by selecting the coded text segment and clicking the *Unmark* button.

Another function on the menu shown in Fig. 8.9 is the **Merge** menu. You can merge similar codes into one by selecting one code first and clicking on *Merge Selected with...* in the **Merge** menu in Fig. 8.9. A new window showing all existing codes will appear; select one or more codes to merge (see Fig. 8.9).

In addition to individually assigning codes to texts in RQDA, users can also employ the R syntax to perform *auto-coding* of a specific word using *pattern matching*. This function allows users to apply a code to paragraphs in a file based on certain words. For example, we wanted to code paragraphs with the keyword *digital*, so we first created a new code called *Digital technology* (see Fig. 8.10).

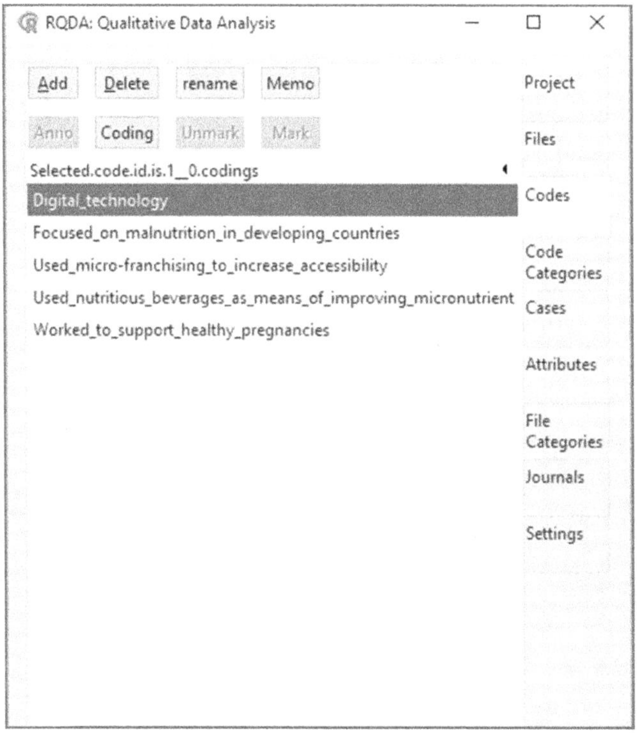

Fig. 8.10 A newly created Code in RQDA: 'Digital technology'

To better demonstrate the *auto-coding* function, we did not manually mark any text segments under this code. To command R to auto code, type the following syntax in the R console:

```
codingBySearch("digital",fid=1,cid=6)
codingBySearch("digital",fid=getFileIds(),cid=6)
codingBySearch("digital",fid=getFileIds(),cid=6,sepera
tor="[.!?]")
```

In the above syntax, *fid* (meaning file ID) represents your selected file. You may find the ID of a file in the GUI **File** tab under the horizontal menu across the top by clicking on the file (see Fig. 8.11).

For example, if we want to auto-code the file **Apart of Me_non-winners**, input 1 after *fid* = since its file ID is 1. Similarly, input *cid* (meaning code ID) of the selected code that you want to auto-coding. In our demonstration, the code ID of the newly created code *Digital technology* is 6 (see Fig. 8.10), therefore we should put a number 6 after 'cid='. After we entered the syntax into the R console and

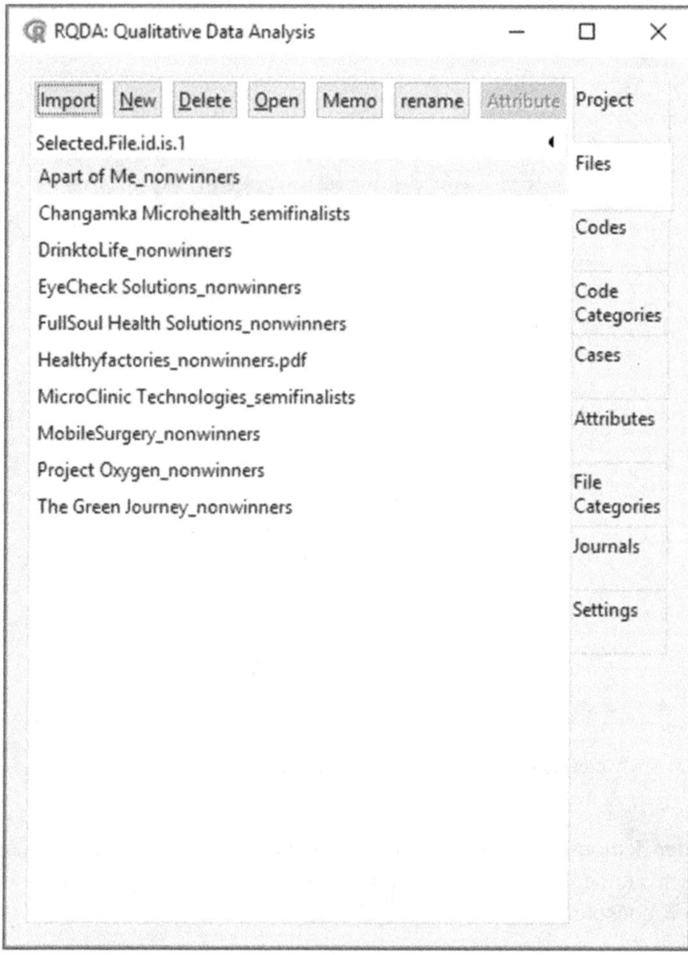

Fig. 8.11 File ID of a file in RQDA (apart of Me_non-winners)

pressed Enter to run the syntax, the auto-coding will complete in a few seconds and you may review the new coding by double-clicking on the code *Digital_technology*. As you can see in Fig. 8.12, each coding contains a keyword '*digital*' as we input into R syntax. If you have any problems with or queries about the auto-coding function, type the syntax: *?codingBySearch* in an R console to get more information.

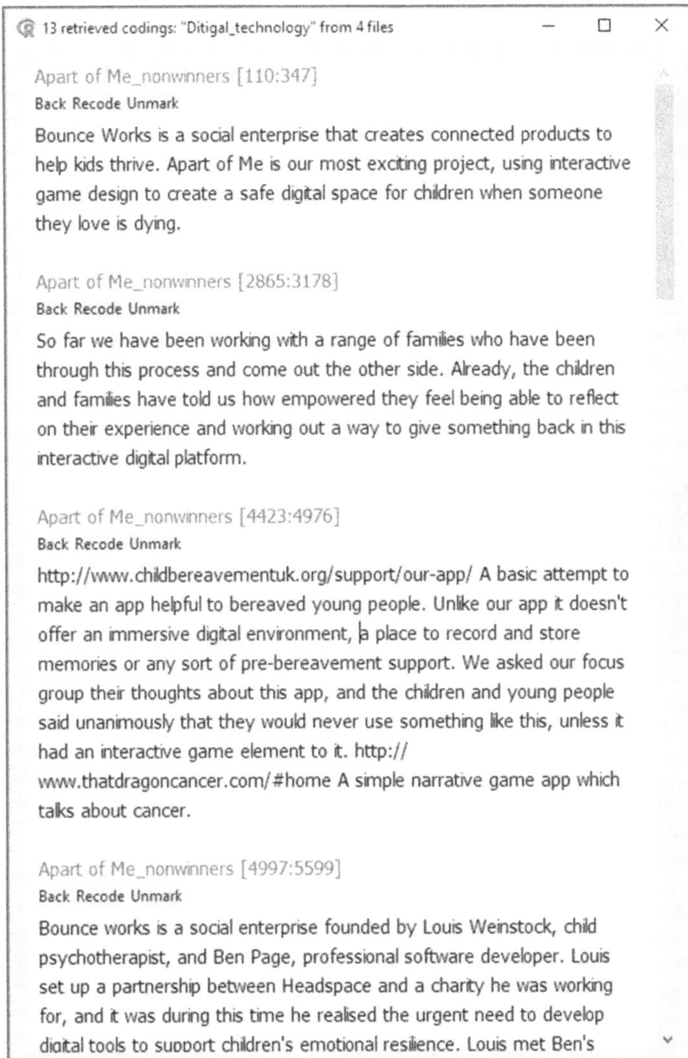

Fig. 8.12 Auto-codings in RQDA

References

Bazeley, P., & Jackson, K. (Eds.). (2013). *Qualitative data analysis with NVivo*. London: Sage.

Boyatzis, R. E. (1998). *Transforming qualitative information: Thematic analysis and code development*. London: Sage.

Corbin, J., & Strauss, A. (1990). Grounded theory research: Procedures, canons and evaluative criteria. *Zeitschrift für Soziologie, 19*(6), 418–427.

Fereday, J., & Muir-Cochrane, E. (2006). Demonstrating rigor using thematic analysis: A hybrid approach of inductive and deductive coding and theme development. *International Journal of Qualitative Methods, 5*(1), 80–92.

Gioia, D. A., Corley, K. G., & Hamilton, A. L. (2013). Seeking qualitative rigor in inductive research: Notes on the Gioia methodology. *Organizational Research Methods, 16*(1), 15–31.

Scheunemann, L. P., Cunningham, T. V., Arnold, R. M., Buddadhumaruk, P., & White, D. B. (2015). How clinicians discuss critically ill patients' preferences and values with surrogates: An empirical analysis. *Critical Care Medicine, 43*(4), 757.

Strauss, A., & Corbin, J. (1998). *Basics of qualitative research techniques*. London: Sage.

Thomas, D. R. (2006). A general inductive approach for analyzing qualitative evaluation data. *American Journal of Evaluation, 27*(2), 237–246.

Verhees, F. J., & Meulenberg, M. T. (2004). Market orientation, innovativeness, product innovation, and performance in small firms. *Journal of Small Business Management, 42*(2), 134–154.

Chapter 9
Data Attributes and Memos

Abstract This chapter discusses techniques to manage files by assigning different case attributes, creating memos or aggregating multiple text files as a single **case** as a way of supporting your qualitative data analysis. RQDA enables users to manage files by assigning attributes, writing memos or aggregating multiple files into a single **case**. These two functions enable researchers both to assign variables to each file and perform further qualitative comparative studies.

Keywords File attributes · Cases · Case attributes · File categories · Memos · Journals

9.1 Creating Data File Attributes

The **Attribute** menu enables you to add variables to individual data/text files. An attribute can be understood as a variable attached to a case or a file with a single value. Attributes can include the type of files (e.g., biographies or interviews) or more detailed information (e.g., on non-winners versus the semi-finalists). Other examples of an attribute include the gender (whether an interviewee is male or female), age (adult or elderly), location (urban or rural area), or performance (high or low performance). The attribute data enables you to perform some simple statistical analysis on the data as additional analysis (after the main data analysis is completed).

The advantage of the **Attribute** function in RQDA is that you cannot inadvertently assign multiple values to a file or a case that should be mutually exclusive. For example, an interviewee cannot be given an attribute of winners and non-winners simultaneously; it has to be either one. To assign an attribute to a file, take the following steps. First, you should go to the **Attributes** menu and click the **Add** button. A new input window will pop up (see Fig. 9.1); next enter a name for the new attribute.

For example, in our illustrative project, "Making More Health: Co-Creating a Healthier World" (or MMH), we extracted profiles from social innovator groups, non-winners, and semi-finalists. To indicate such groupings, we created an attribute

Y. Chandra, L. Shang, *Qualitative Research Using R: A Systematic Approach*,
https://doi.org/10.1007/978-981-13-3170-1_9

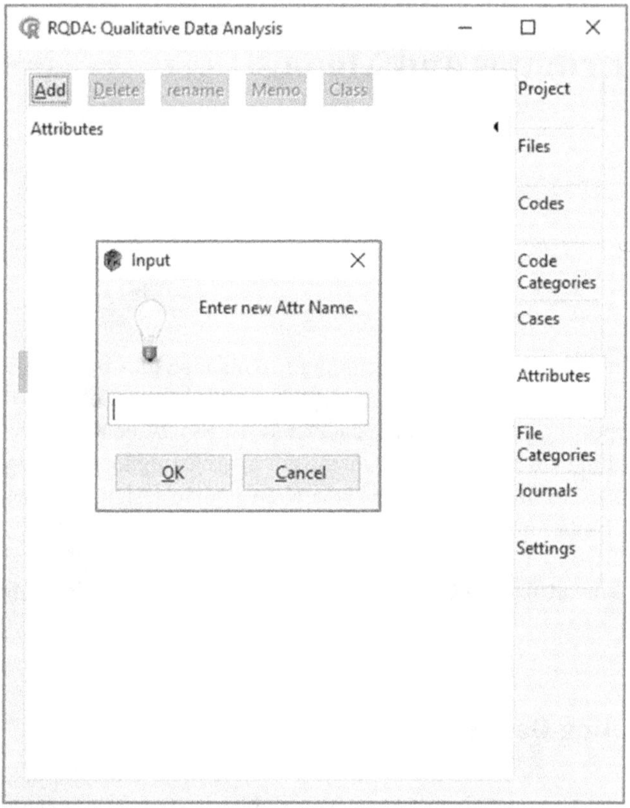

Fig. 9.1 Adding data attributes in RQDA

called *Competition Results*. We also created two attributes, *Organization Type* and *Project Stage*. The created attributes can be found under the **Attributes** tab in the menu at the top of the page, as shown in Fig. 9.2.

RQDA has two types of attributes: numbers and letters. For each attribute, you can assign a type by using the **Class** function. To do so, click on the **Class** button in the **Attributes** tab and a new window with three choices will appear (see Fig. 9.3).

If you select 'numeric' for the type of attribute, then you can use numbers to assign variables that represent different categories under each attribute. For example, we chose *numeric* for the attribute *Competition Results* and we decided to use '0' to represent the non-winners' and '1' to represent the semi-finalists' group. To do this, first select the attribute 'Competition Results' and then click on the '*Memo*' button in the **Attributes** tab, and a notepad window will pop up. Then assign the variables in the notepad using the format shown in Fig. 9.4.

By doing so, we have successfully assigned two variables for the attribute '*Competition Results*'. Remember to save the memo by clicking on the '**Save**

Fig. 9.2 Viewing existing attributes in RQDA

Fig. 9.3 Selecting type of attribute in RQDA

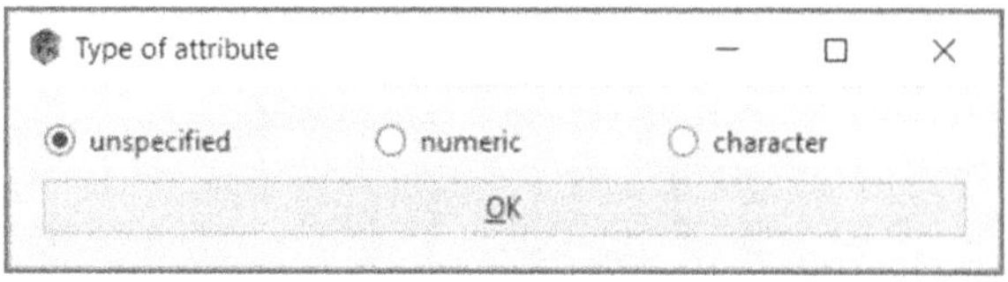

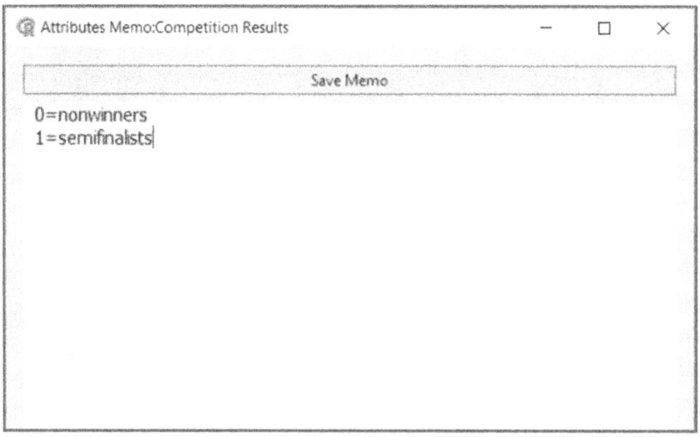

Fig. 9.4 Assigning attribute variables in RQDA

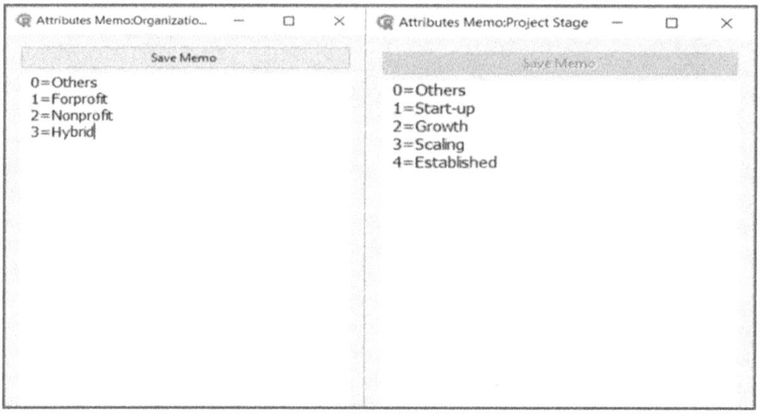

Fig. 9.5 Viewing attributes variables in RQDA

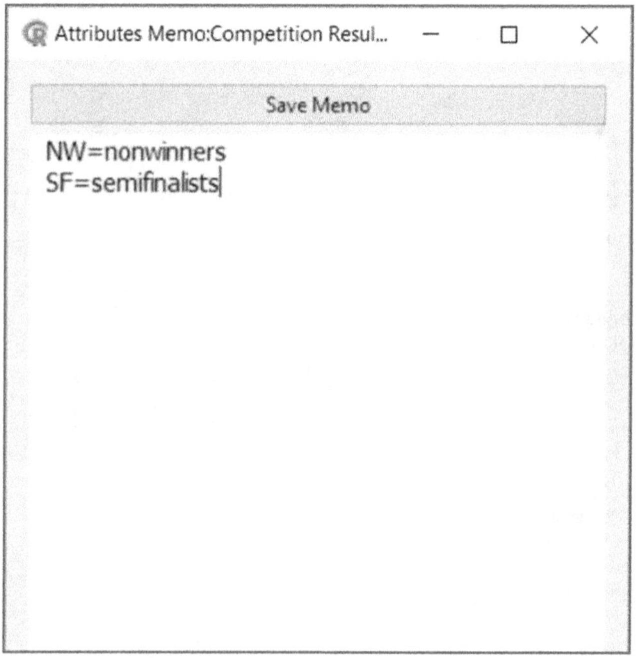

Fig. 9.6 Character type of attribute in RQDA

Memo' button before you close the window. Similarly, we used the same strategy to assign variables for the other two attributes respectively (see Fig. 9.5).

If you choose **'character'** type for the attribute, then you must use words/characters to represent attribute variables. For example, we can use 'NW' and 'SF' to represent the *non-winners'* group and *semi-finalists'* group respectively (see Fig. 9.6).

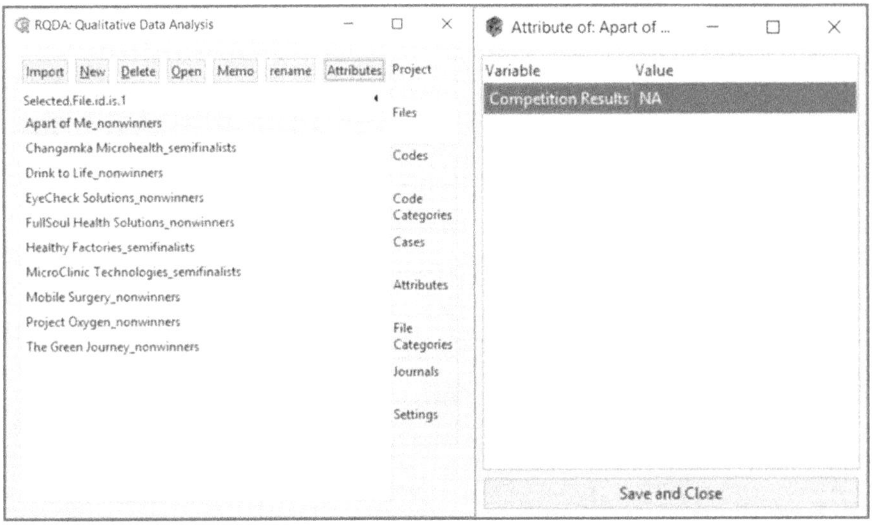

Fig. 9.7 Assigning attributes to a file in RQDA

After completing the attributes setting, assign the attributes to each of the imported files. You first need to select files in the **Files** tab and then click the Attribute from the top menu. A new window with attribute information will pop up as shown in Fig. 9.7.

To assign attributes, click '**NA**' under the **Value** section in the pop-up window and enter the attribute value. For example, the social innovation project 'Apart of Me' in our illustrative project did not win the Changemaker competition, and thus we entered '0' in the value section, indicating that the file of the 'Apart of Me' case belongs to the non-winners' group. Click on the blank space after entering the value and then click the '**Save and Close**' button to save the changes, and the attribute value will appear in the pop-up window (see Fig. 9.8).

To review all attributes, right-click on the blank space in the **Files** tab and select **View Attributes** in the pop-up menu. A new window with the attributes of all files will appear, which offers a clear overview of the files. For example, as Fig. 9.9 shows, our sample of social innovation projects include four for-profit organizations, four hybrid (i.e., profit and nonprofit) organizations, and one non-profit organization.

9.2 Creating Cases

In the **Cases** menu, users can associate several files with a particular case (which represents units of analysis; you can link a case to a subset of files), which enables users to cull files with similar characteristics (e.g., all text files involving *Ashoka Fellows from India*) and compare them with other groups of files (e.g., all *Ashoka*

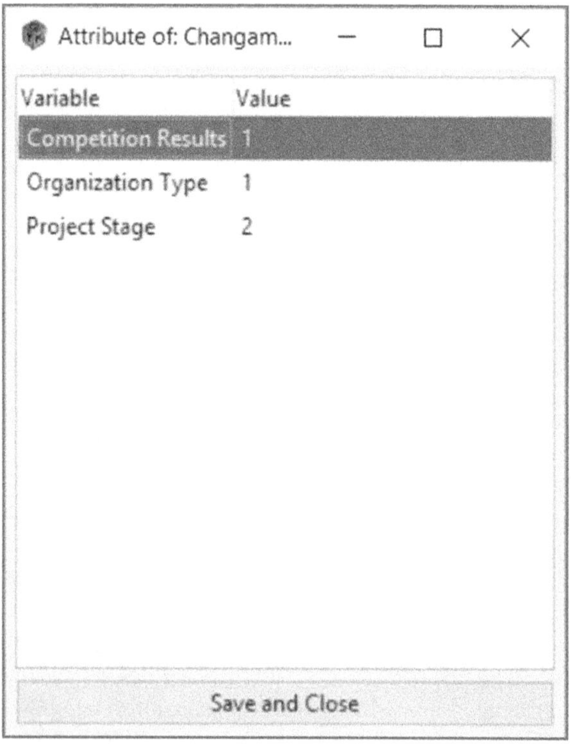

Fig. 9.8 Viewing file attributes in RQDA

Fellows from the US). Creating cases is quite simple: simply click the **Add** button in the **Cases** menu and enter a name for the new case (see Fig. 9.10 below).

To demonstrate this, we created two cases: *developed country* and *developing country*. We use these two cases to show the country of origin of social innovation projects. Users can add files to the cases by taking the following steps illustrated in Fig. 9.11.

1. Right-click on the case (e.g., *developing country*) in the **Cases** tab to which you want to add a file.
2. Click **Add File(s)**.
3. Choose the file(s) that you want to add from the pop-up window.

An alternative way to add files to cases is to right-click on the file in **Files** tab, and select **Add To Case…** in the pop-up menu. Then click on the case or cases that you want to add the file from the new pop-up window (see Fig. 9.12).

After you successfully add files to cases, you can view all files associated with each case found in the **Cases** tab by clicking on the case name below the top menu. For instance, to view which social innovation projects' *regions of impacts* are in developing countries, we simply click on the case **Developing country** and a list

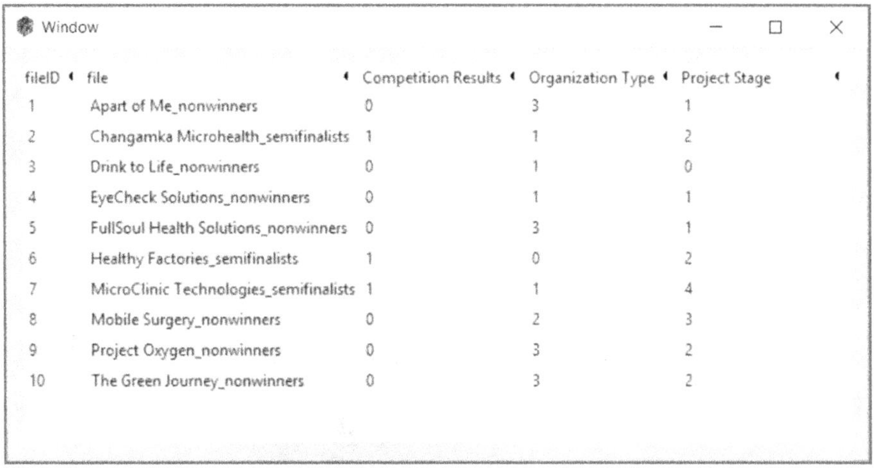

Window

fileID	file	Competition Results	Organization Type	Project Stage
1	Apart of Me_nonwinners	0	3	1
2	Changamka Microhealth_semifinalists	1	1	2
3	Drink to Life_nonwinners	0	1	0
4	EyeCheck Solutions_nonwinners	0	1	1
5	FullSoul Health Solutions_nonwinners	0	3	1
6	Healthy Factories_semifinalists	1	0	2
7	MicroClinic Technologies_semifinalists	1	1	4
8	Mobile Surgery_nonwinners	0	2	3
9	Project Oxygen_nonwinners	0	3	2
10	The Green Journey_nonwinners	0	3	2

Fig. 9.9 Reviewing data attributes in RQDA

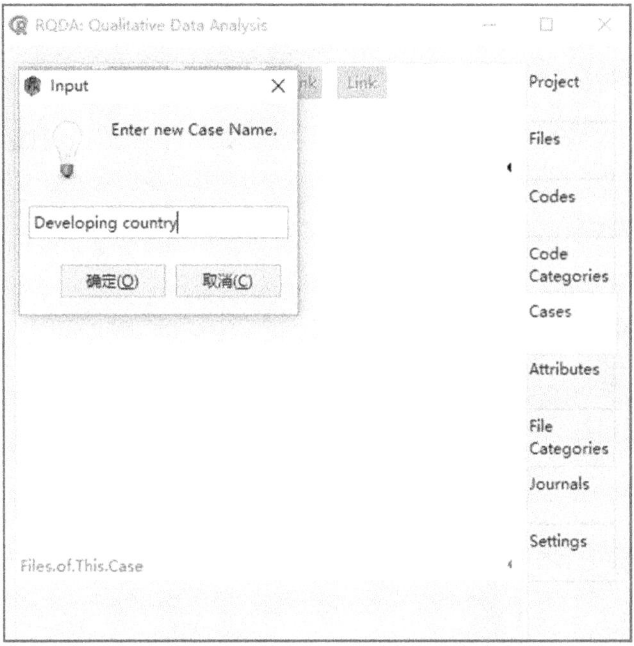

Fig. 9.10 Creating new cases in RQDA

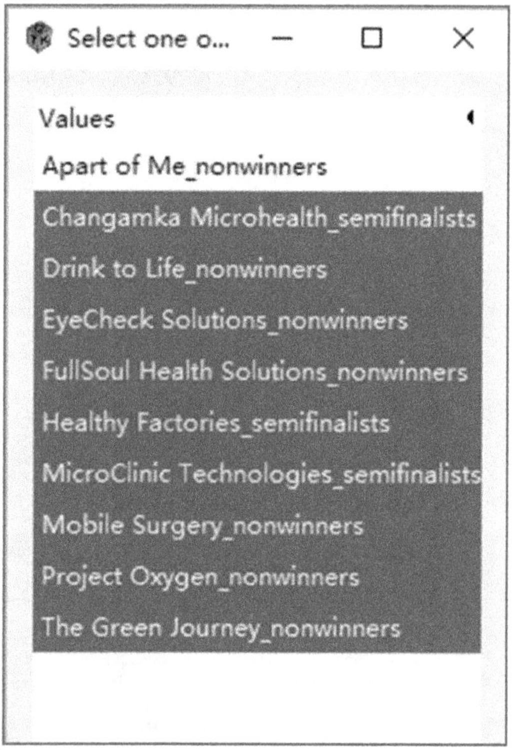

Fig. 9.11 Adding files to cases in RQDA

Fig. 9.12 Adding individual file to case/cases in RQDA

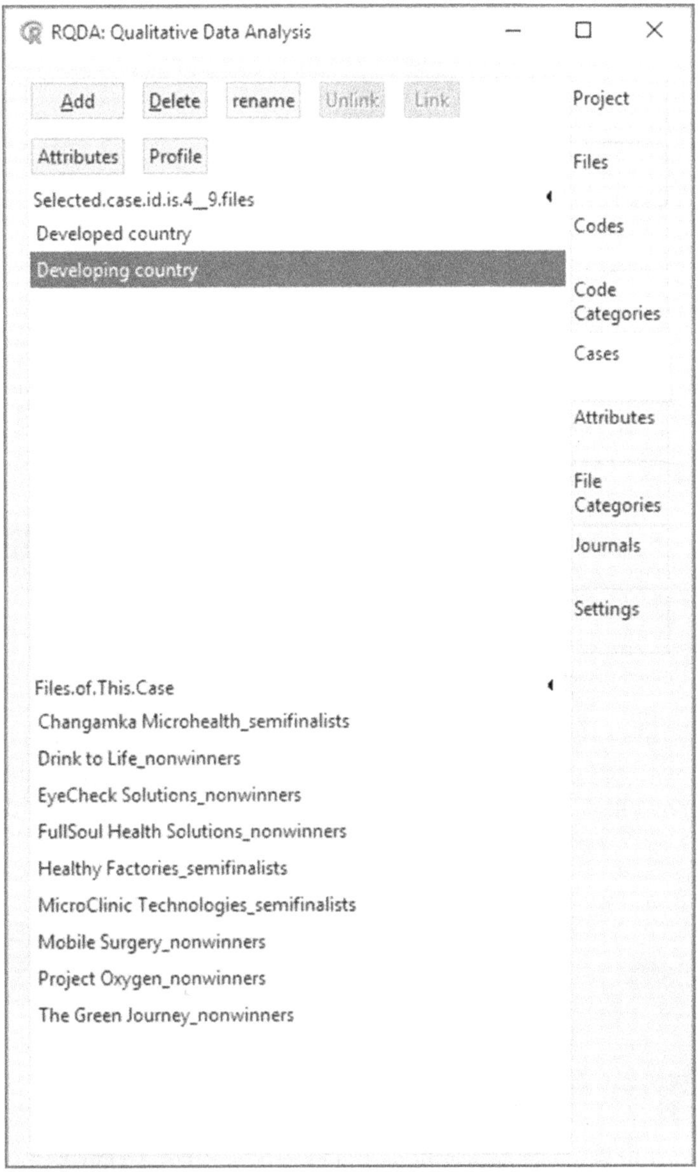

Fig. 9.13 Viewing files in the 'Developing Country' case in RQDA

of files associated this case will appear in the bottom window (as shown in Fig. 9.13).

One useful feature of the **Case** function is that it allows users to link or unlink part(s) of the file to or from, respectively, cases. For example, we can link just the first paragraph of the *Apart of Me* file (see Fig. 9.13) and unlink the remainder of the

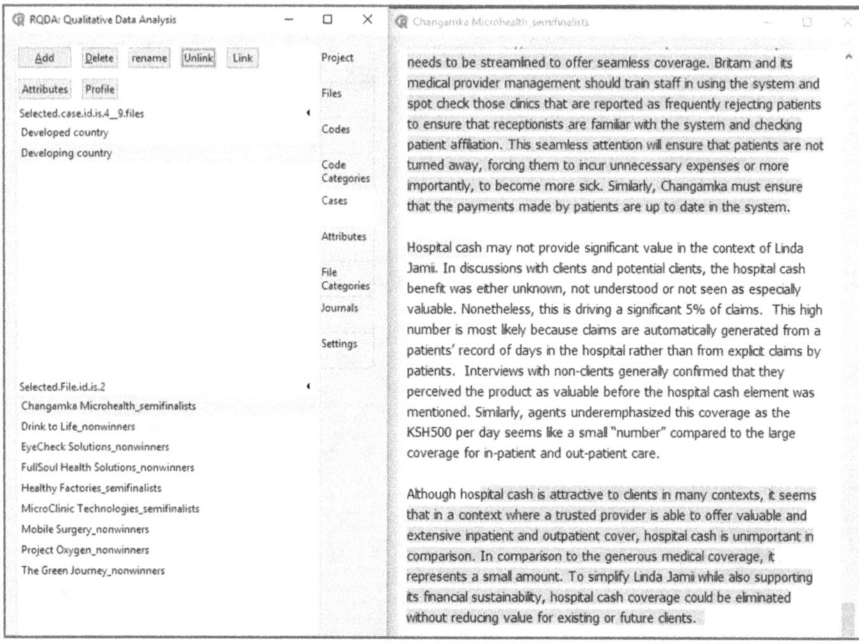

Fig. 9.14 Unlinking partial texts to the case in RQDA

text. To do so, we first open the *Apart of Me* file in the **Cases** menu. Then, we select the text segments we want to remove from the case and then click on the *Unlink* button at the top of the **Cases** menu. The selected text will be removed from the highlights (as shown in Fig. 9.14). This function can be quite useful when we want to focus on some specific parts of texts in the files for analysis. For instance, if our files are from interview transcripts, we may only want to analyze interviewees' responses. We can use this function to only link the responses from different interviews to a case.

9.3 Creating File Categories

Another way to organize files in RQDA is to use the **File Categories** menu. This menu allows users to classify files into various categories. For example, when we analyzed our sample project, we wanted to classify the files according to the data sources. Unlike the **Attribute** function, users can assign multiple categories to a file. For demonstration, we created two file categories, *sole source* and *multiple*

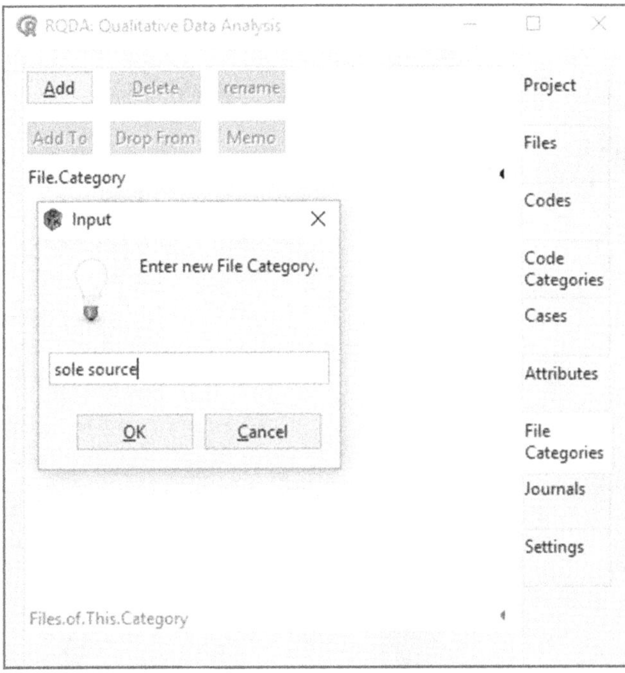

Fig. 9.15 Adding new file categories in RQDA

sources, in the File Categories menu by clicking the **Add** button as shown in Fig. 9.15.

To add files to a file category, click the selected *file category* in the **File Categories** tab, and then click on the **Add To** button, and select one or more files you want to add from the new pop-up window (see Fig. 9.16).

Alternatively, files may also be added into File categories by pointing to individual files, right-clicking on selected file(s) from the **Files** menu and then selecting *Add To File Category...* from the pop up menu. Next, users can select one or more file categories and click **OK**. This alternative enables users to simultaneously add a file(s) into multiple categories (see Fig. 9.17 below).

However, if users want to add a batch of files to a single file category, it is more convenient to follow the first option discussed above. Once all the steps are successfully executed, the added files will appear in the **File Categories** menu under different categories, as shown in Fig. 9.18. For instance, Fig. 9.18 shows that the Ashoka file category consists of 7 Fellows (from ‘Drink to Life’ to ‘The Green Journey’).

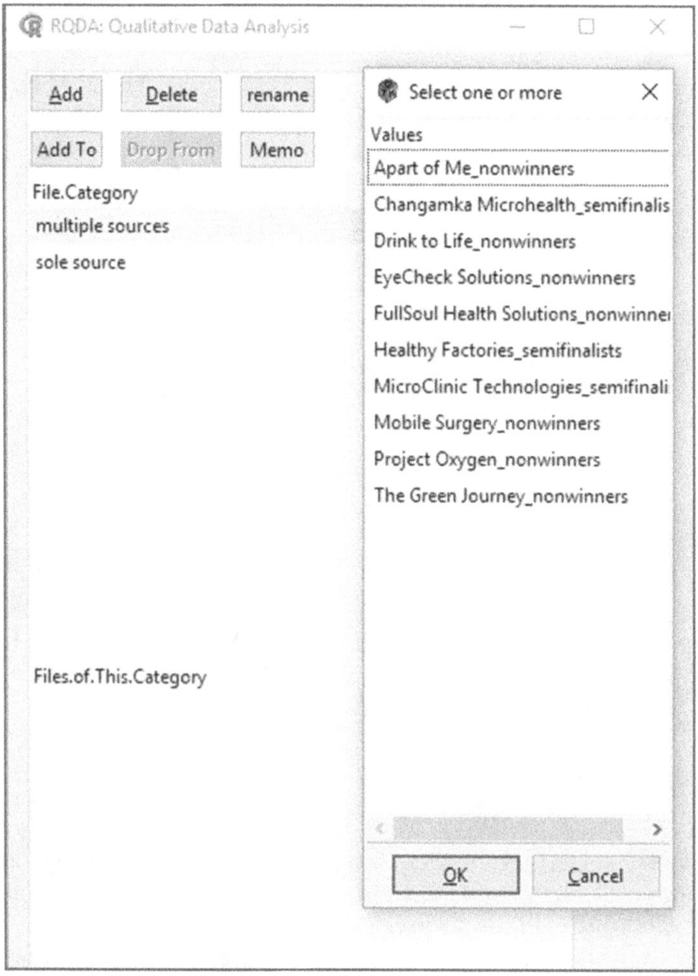

Fig. 9.16 Adding files to a file category in RQDA

You can find specific files within a single file category, by using keywords. Under the **Files Categories** menu, simply right-click on the blank space under the *Selected.file.id.is.XX* section and select *Search Files Within Categories* and enter keywords into the pop-up input window to begin the search. For example, if we want to search social innovation projects' files that contain keyword '*innovation*', we type *file like '%innovation%'* in the pop-up input window (see Fig. 9.19).

After clicking *OK*, we can see that only four files are in this category, which means that only these four files contain the keyword *innovation* (see Fig. 9.20).

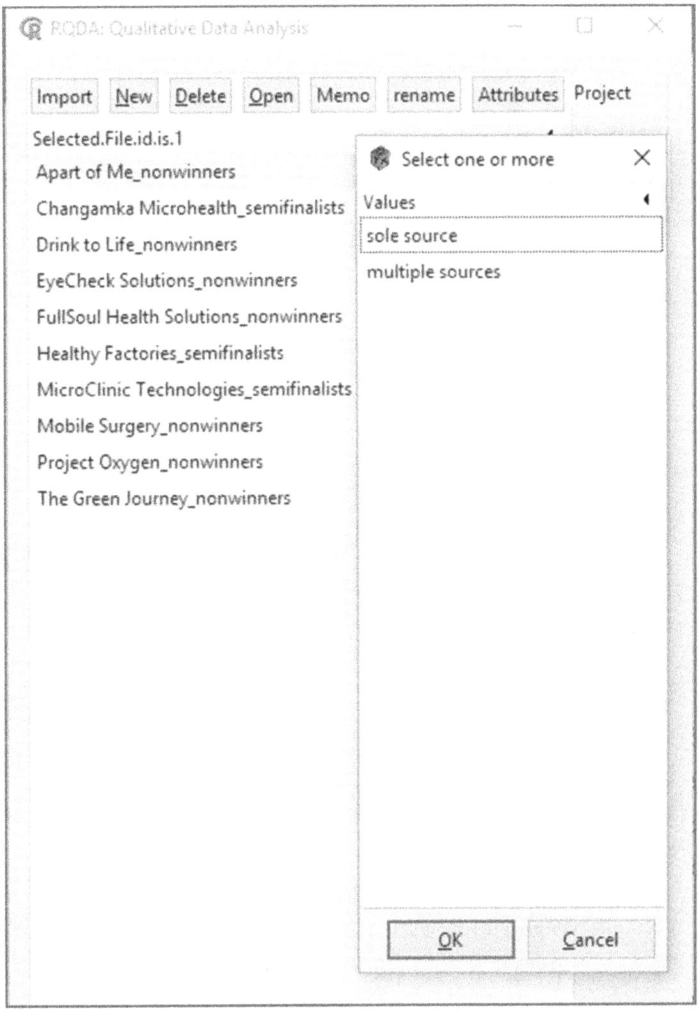

Fig. 9.17 An alternative way to add a file(s) to a file category in RQDA

9.4 Writing Memos and Journals

Users can record their thoughts and important ideas, decisions or actions during the coding process by using the **memo** and **journal** tools. These tools help researchers organize data and record important codes, or questions about codes, thus facilitating the data aggregation and/or theory-building processes. In particular, RQDA allows users to create different *levels* of memos. For example, users can create a *general* memo for an entire project simply by clicking the **Project Memo** button in the **Project** tab, entering the memo content and clicking the **Save Memo** button at the

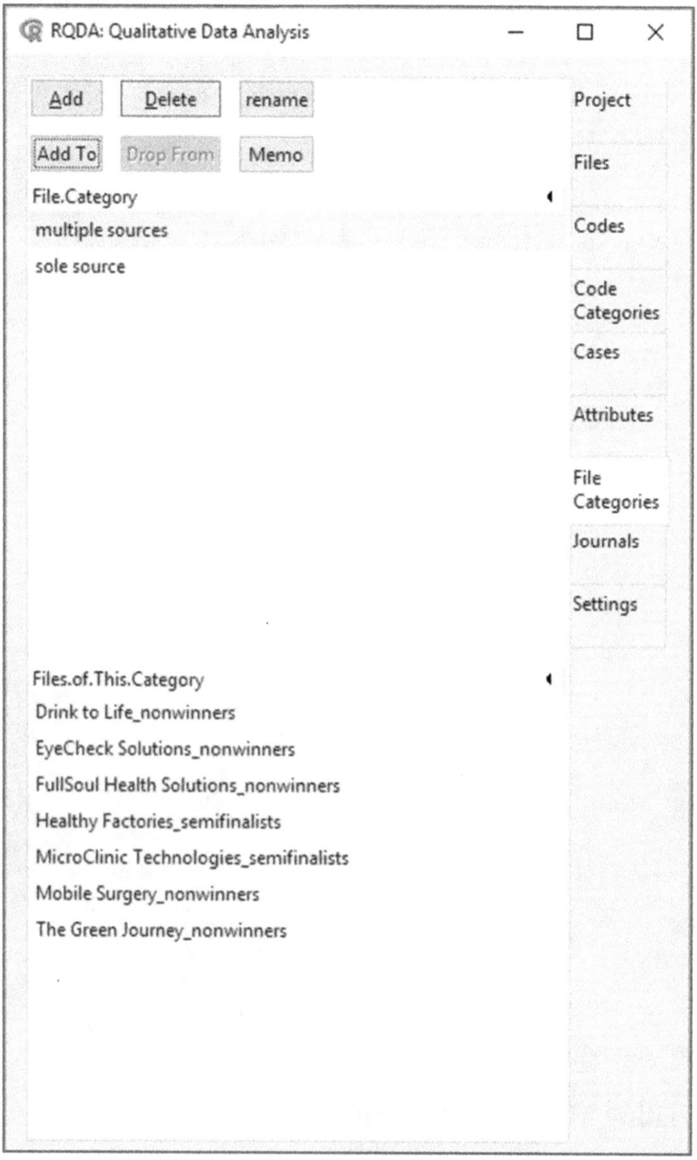

Fig. 9.18 Viewing files under a file category in RQDA

top of the window (see Fig. 9.21). You can also attach memos to *specific* items, such as individual codes, files, file categories and cases.

The **Journal** function allows users to record their thoughts during the coding process (For instance, social innovation project #1 Apart of Me is very innovative in terms of their new products and new business model; project #2 Changamka is unusual because it does not have A, B and C elements usually found in a social

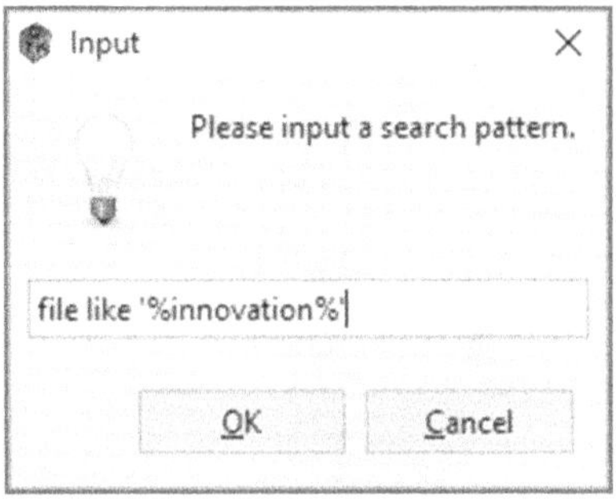

Fig. 9.19 Searching for files under a file category with a keyword

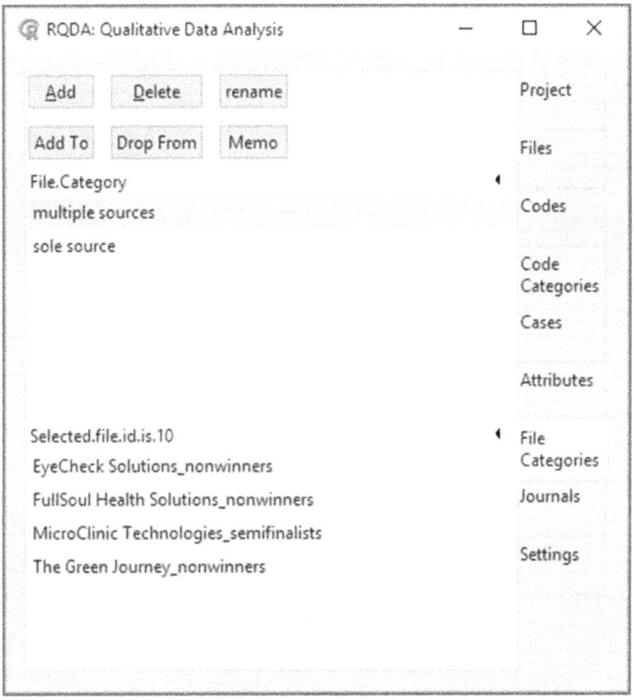

Fig. 9.20 Viewing files with the keyword 'Innovation' in RQDA

innovation). Creating a new journal is similar to creating new cases or file categories: users simply click the **Add** button in the **Journals** tab and a new input window will appear (see Fig. 9.22).

Journals can be extremely useful in managing large-scale qualitative research projects. For example, if you have hundreds or thousands of case files, you will need to do the coding and journaling across an extended period of time, possibly up to a few months, so keeping a journal of your thoughts, questions and problems can help you work through different issues you may encounter. By using the journal func-

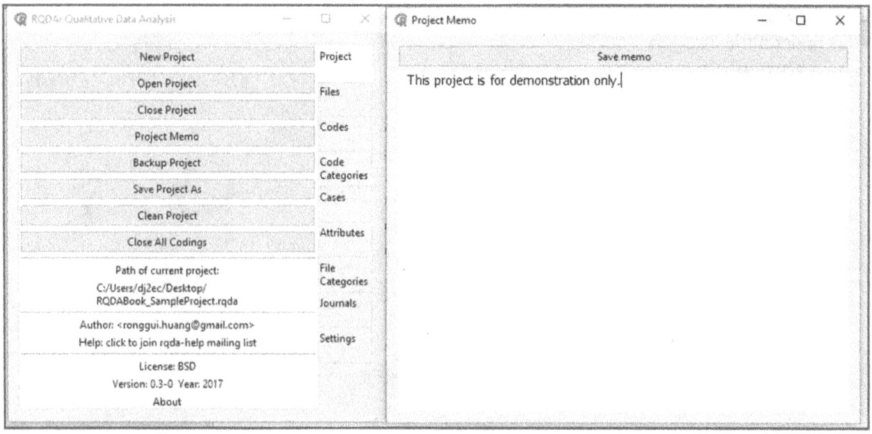

Fig. 9.21 Adding a memo to the project in the RQDA

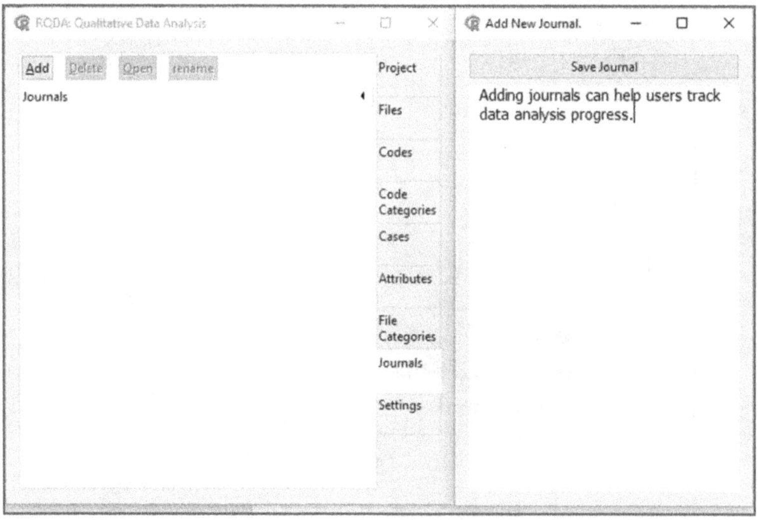

Fig. 9.22 Creating journals in RQDA

tion, users can write down information about the progress of the project (e.g., "as of this date and hour, I stopped coding up to here; I have not yet done X, Y and Z") so that when they continue working on the project, the journal can help them track the project's progress or deal with problems that they encounter or anticipate, or make changes in an approach to analyzing the data. RQDA also shows the time at which each journal was created within the **Journals** menu to assist in good housekeeping.

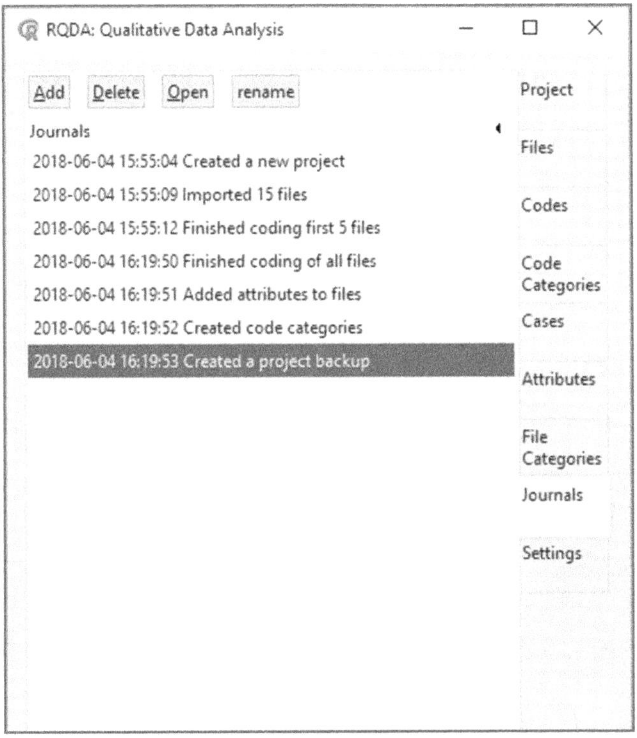

Fig. 9.23 A screenshot of multiple journal entries in RQDA

Unlike the **Memo** function where users could add memos to a case, a code, or even a project, the **Journal** function does not limit the notes to a specific type, therefore, there is no limitation on the number of journal entries that can be created or written. For instance, users can create different journal entries to keep a record of data analysis process (see Fig. 9.23. for an example).

Chapter 10
Codes Aggregation, Plotting and Exporting

Abstract After completing the coding process, the next step is to organize the codes. This chapter discusses the techniques used in aggregating codes in RQDA, plotting the codes in a two-dimensional network format, and exporting the plotted networks via an HTML file that is shared with fellow collaborators and journal gatekeepers.

Keywords Codes aggregation · Code categories · Plotting · Exporting · Two-dimensional network

10.1 Aggregating and Plotting Code Categories

After coding all the first-order codes (or open coding) for all the textual data imported into RQDA, the first-order codes can be abstracted or merged into higher code categories (or into *second-order themes*) to facilitate building the grounded theory process, which is accomplished using the **Code Categories** menu. Aggregating the *first*-order codes to *second*-order codes or themes (or axial coding) is an important step because researchers not only begin to firmly assign code labels based on similarities and differences among the many first-order code categories but also seek to identify relationships between and among first- and second-order concepts (e.g. first-order codes "Developed_low-costs_childbirth_equipment" and "Built_affordable_vision_diagnostic_tools" can be abstracted into "Increasing affordability of health services" as a second-order theme)(Gioia et al. 2013; Strauss and Corbin 1998; Tracey et al. 2011). This process is essentially *iterative*, and users can amend the first-order codes and their associated assignment to the second-order themes. To create a second-order theme that consists of multiple first-order codes, users must click the **Add** button under the **Code Categories** menu and enter a name for the new code category (see Fig. 10.1).

The **Code Categories** function similarly to that of the **File Categories**. Users can categorize codes to **Code Categories** as the second-order themes. When the coding aggregation to the second-order themes is finished, the aggregated second-

Y. Chandra, L. Shang, *Qualitative Research Using R: A Systematic Approach*,
https://doi.org/10.1007/978-981-13-3170-1_10

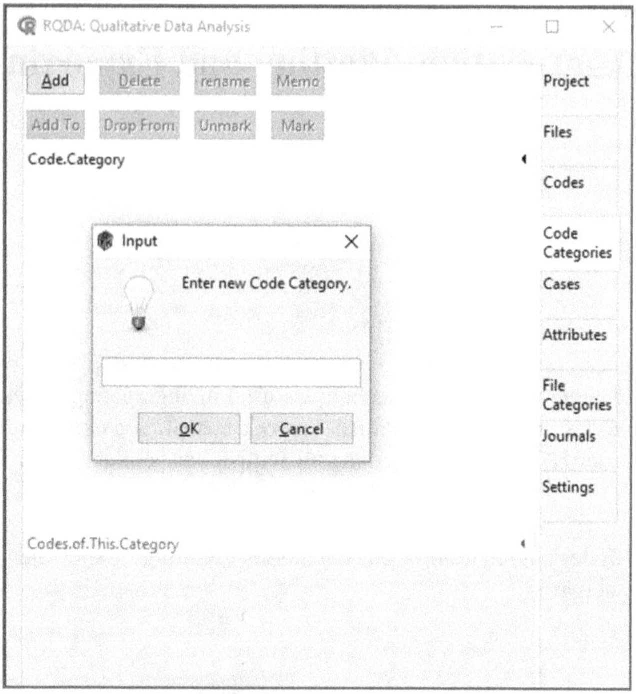

Fig. 10.1 Adding code category in RQDA

order codes or themes need to be reviewed and the potential relationships among, explored. For example, in our sample project, "Making More Health: Co-Creating a Healthier World" (or MMH), we identified four first-order codes related to the development of new products: (1) '*Develop a health management information software to track stockpile of commodities*', (2) '*Develop vision care tools that bring objectivity, speed and mobility*', (3) '*Develop field-based health education program*' and (4) '*Develop naturally fortified foods to prevent malnutrition and obesity*'. Next, we created a second-order code category **'New products development'** in the **Code Categories** menu to represent and summarize the three first-order codes, as shown in Fig. 10.2.

Next, we added the four first-level codes (mentioned in the previous paragraph) to the **'New products development'** category simply by clicking the **Add To** button from the top menu and selecting the codes that we want to add from the pop-up window (see Fig. 10.3 below).

After adding new first-order codes to the second-order themes (or second-order code categories), you can review a second-order theme by clicking it in the Code Categories tab, and by doing so, all the first-order codes that belong to that theme will appear in the window (see Fig. 10.4).

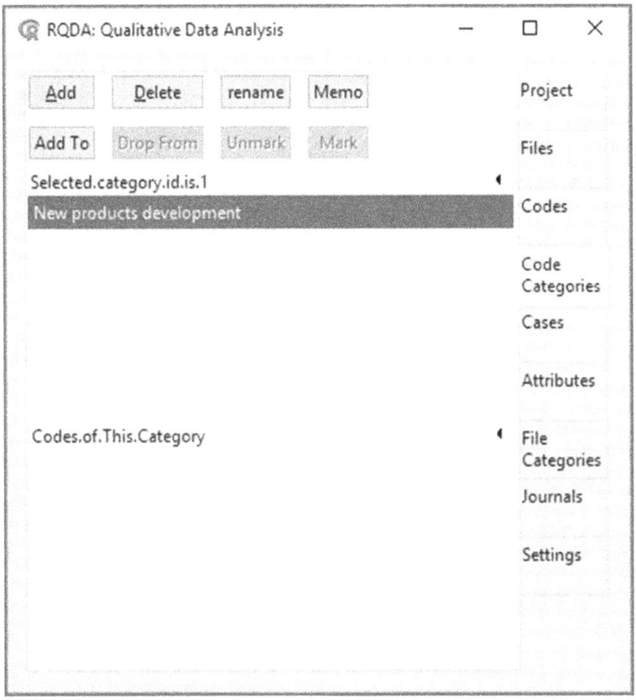

Fig. 10.2 A code category in RQDA

In RQDA, users can create a new first-order code and add to the selected second-order theme directly within the **Code Categories** menu. To do this, right-click on the selected (second-order theme) category and select **Add New Code to Selected Category** in the new pop-up menu. Next, enter a new second-order code in the pop-up text-entry window as shown in Fig. 10.5.

To retrieve and review all the codes of second-order theme categories, right-click on the selected code category and select *Codings of selected category* in the pop-up menu, and then a new window showing all codes associated with the code category will appear (see Fig. 10.6).

When the code aggregation is complete, users can plot the coding results in a two-dimensional output. This process produces *nodes* and *arrows* in the form of networks in which each node represents first-order codes connected to second-order themes at the center of the network. The output can be modified using various network types (e.g., Fruchterman-Reingold or Kamada-Kawai layout) and is visually appealing. Users can plot all second-order themes from the **Code Categories** menu by right-clicking on the selected categorie(s) and selecting *Plot Selected Code Categories*. If you select *Plot Selected Code Categories*, a new window will pop up showing the selected code-category relationships. A first-order

Fig. 10.3 Adding codes to a code category in RQDA

code may be aggregated into more than one second-level code category (i.e., multiple coding for each text segment) and such connections will also be shown in the plot map (see Fig. 10.7). This plot map shows that, for example, a second-order theme called "***Empowering patients***" comprises four first-order codes: "*empowered children to handle negative emotions*", "*formed peer support groups for patients*", "*used interactive games to enhance children's emotion and wellbeing*", and "*facilitate human interaction and communication through an app*" (see the codes at the bottom of Fig. 10.7).

The layout of the plot map can be modified for optimal viewing by clicking the **Layout** button from the top menu; a pop-up menu will appear. From the **View** menu, drag the individual first-order codes or second-order code categories to adjust their position, and rotate the plot map around the origin by different degrees (e.g., at 15, 45 and 90 degrees clockwise and counter-clockwise) to better view the plot map. Besides plotting all second-order codes, you may also plot just a portion of them by selecting only a few second-order codes that you want to plot and better see their relationship to their respective first-order codes. For example, in our sample project, we wanted to see the connections between two second-order codes '*Using informa-*

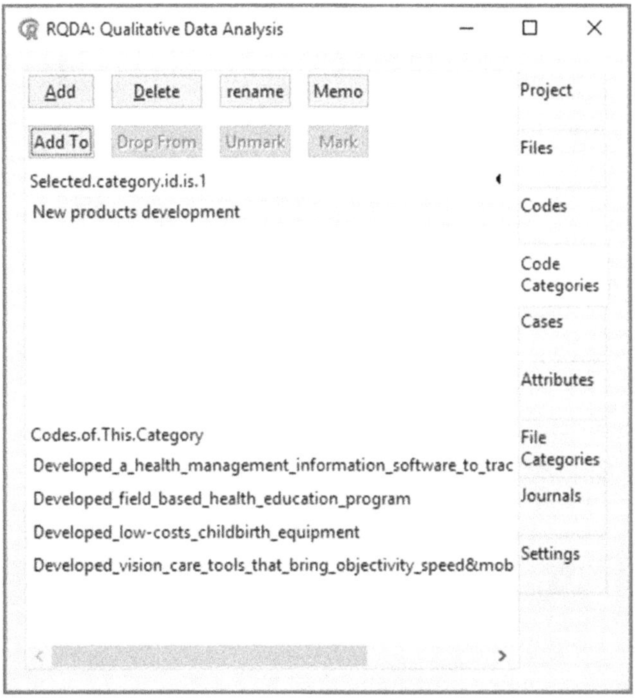

Fig. 10.4 Reviewing codes of a code category in RQDA

tion communication technology' and '*Increasing healthcare affordability*', so we selected these two second-order codes in the Code Categories tab and plotted the output (see Fig. 10.8). From the figure you may see that the two second-order Code Categories (*Increasing healthcare affordability &Using information communication technology*) are connected by two common codes: "*Designed_a_free_software_for_children*" and "*Created_a_mobile_phone_platform_for_health_insurance*". Such connections can lead to further exploration and discussion on how the technology development can contribute to the affordability of healthcare services.

Additionally, users also can plot the selected second-order codes with d3 (d3 is a JavaScript library for visualizing data with HTML, SVG, and CSS). To do this, users may directly select the code category/categories that you want to plot, right click and select *Plot Selected Code Categories with d3*, a new HTML file will appear with a visualized plot map (as shown in Fig. 10.9).

You can highlight the codes (first-order codes) or code categories (second-order themes) simply by clicking on them. Then, the selected codes or code categories and their nodes will be highlighted in red.

To date, the code-category relationships in RQDA can only be established up to one hierarchy—from first-order codes to second-order themes. According to the

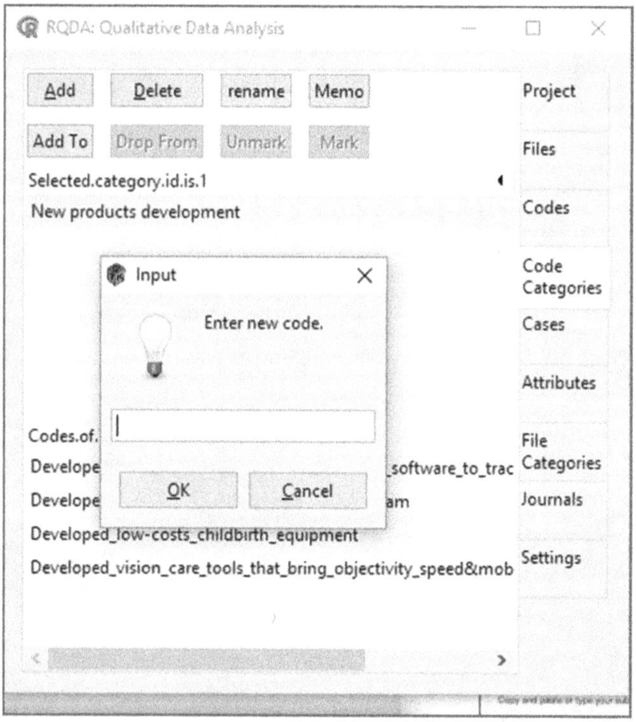

Fig. 10.5 Adding a new code to a selected category in RQDA

developer of RQDA, Ronggui Huang, the original goal in developing RQDA was to simplify the structure of the concepts to clarify the findings (Huang 2014). Therefore, RQDA does not provide a detailed tree-like category structure to show the hierarchy of the concepts. For a process model or a grounded theory project, hierarchies of concepts and higher-level categories must be created in order to show *how* the grounded theory emerges. The above coding samples (Figs. 10.7 and 10.9) look well-ordered up to the second-order code categories because our small demonstration project has relatively few first- and second-order codes (36 first-order codes and 14 second-order themes). But if we have dozens or more cases involving a larger number of text data files, the coding output (the nodes and arrows) can look quite disorganized. Users can create *aggregate dimensions* by creatively exporting their first-order codes and second-order theme categories and using a spreadsheet (e.g., Excel) to iteratively organize all the codes to construct *aggregate theoretical dimensions*.

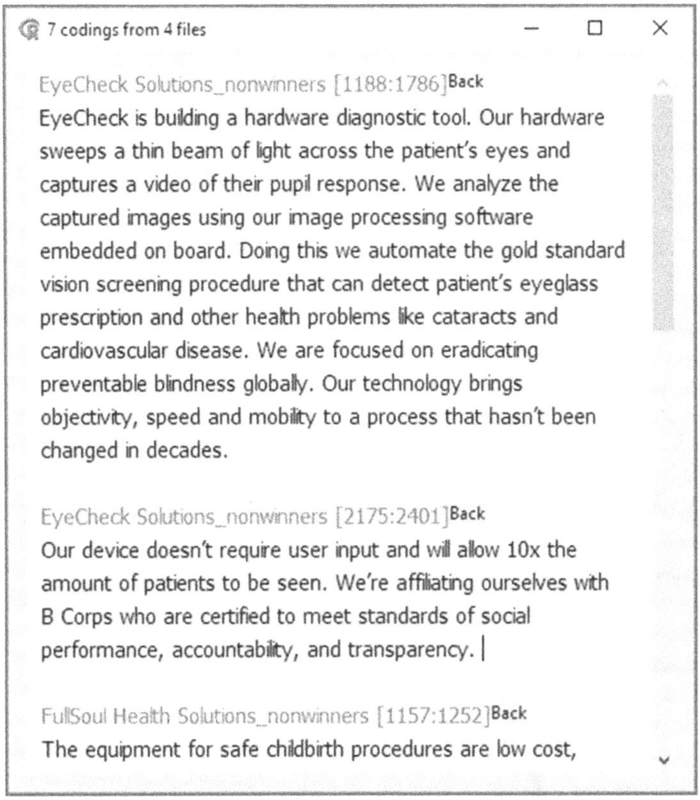

Fig. 10.6 Reviewing codes within a code category in RQDA

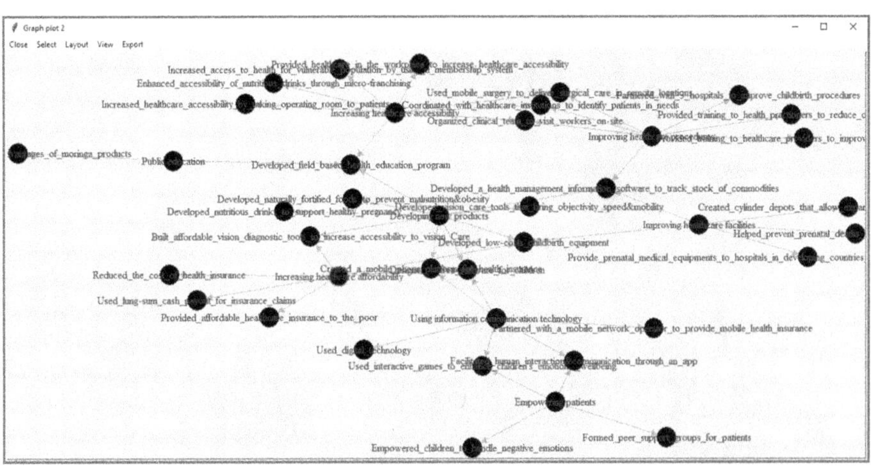

Fig. 10.7 Plotting all code categories in RQDA

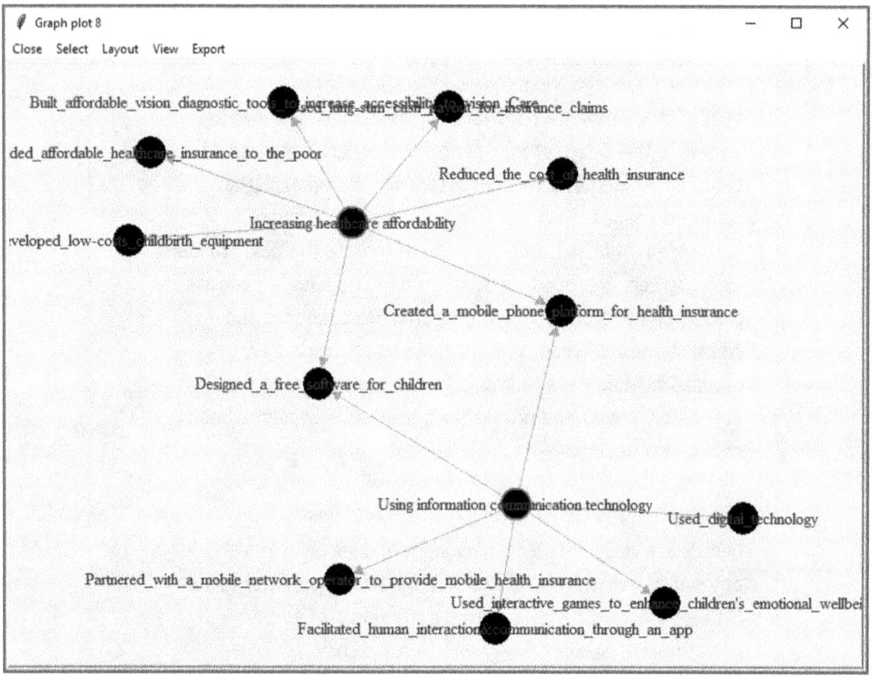

Fig. 10.8 Plotting selected code categories in RQDA

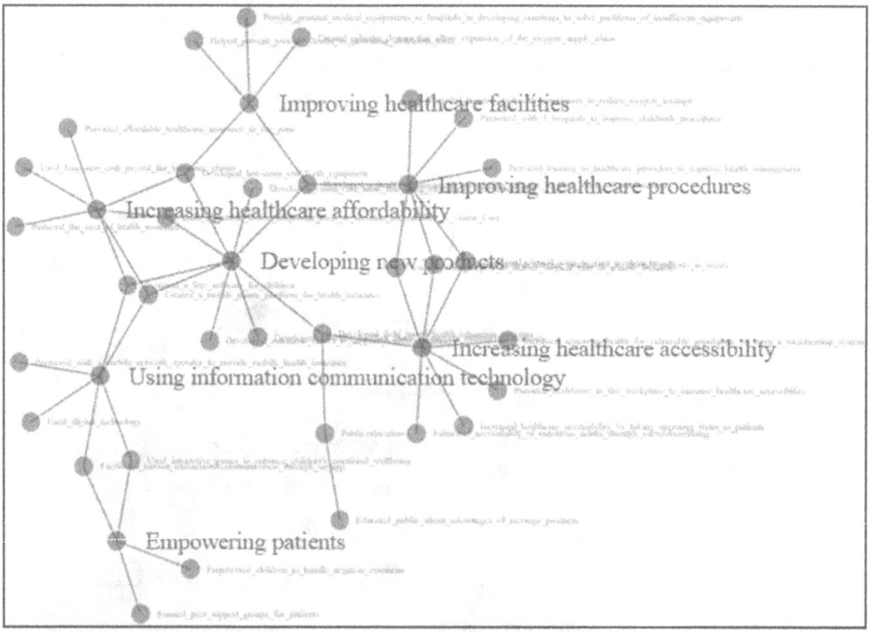

Fig. 10.9 Plotting the code categories with d3 in RQDA

10.2 Exporting Resources

10.2.1 Exporting File or Case Attributes

In RQDA, users can export both **File** attributes and **Case** attributes to facilitate *further qualitative analysis*. To export file attributes, right-click on the blank space within the **Files** tab. Then, you can select *Export File Attributes* from the new pop-up menu (see Fig. 10.10).

A new input window will appear; enter a name for the exported files. The exported attributes will be saved in a new Excel file, as shown in Fig. 10.11.

Similarly, you can also export Case attributes using the same steps by right-clicking on the blank space within the **Cases** tab.

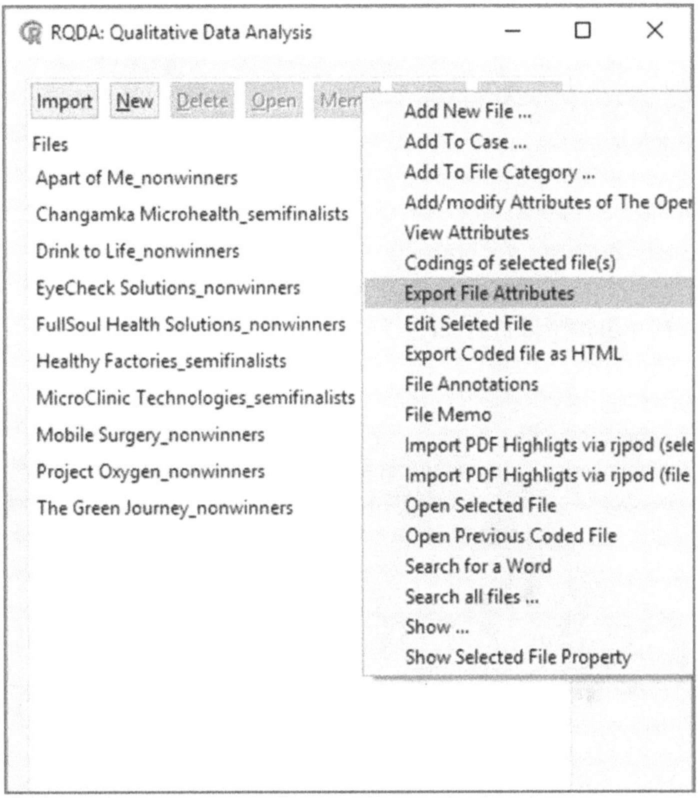

Fig. 10.10 Exporting file attributes in RQDA

fileID	file	Competition Results	Organization Type	Project Stage
1	Apart of Me_nonwinners	0	3	1
2	Changamka Microhealth_semifinalists	1	1	2
3	Drink to Life_nonwinners	0	1	0
4	EyeCheck Solutions_nonwinners	0	1	1
5	FullSoul Health Solutions_nonwinners	0	3	1
6	Healthy Factories_semifinalists	1	0	2
7	MicroClinic Technologies_semifinalists	1	1	4
8	Mobile Surgery_nonwinners	0	2	3
9	Project Oxygen_nonwinners	0	3	2
10	The Green Journey_nonwinners	0	3	2

Fig. 10.11 Exported file attributes of coded files

10.2.2 Exporting Coded Files for Sharing with Research Collaborators and Gatekeepers

In addition to exporting attributes, in RQDA, you can also export coded files to be shared with research collaborators and gatekeepers by right-clicking the selected coded file and then choosing *Export Coded file as HTML.* After entering a new file name, a new HTML file will be created showing all the codes and coded texts of the individual file (see Fig. 10.12).

10.2.3 Exporting Completed Codings

One of the limitations of the "manual" approach (data analysis without using computing resources) in coding qualitative data is the inability to demonstrate *transparency* in the researchers' decisions behind the codings and coding process. To facilitate transparency and trustworthiness, RQDA enables users to export all the completed codings in a project into a single HTML file, which can then be shared with research collaborators and gatekeepers via emails and posted in online data platforms (e.g., Harvard DataVerse or GitHub). All the RQDA codings can be exported by right-clicking the blank space in the Codes tab and selecting *Export Codings as HTML.* A new pop-up window will appear, and you may select all or part of the codes created for exporting. By following these steps, a new HTML file will be created in a target location (see Fig. 10.13 below).

The new HTML file shows a list of exported first-order codes (left screen) and the quotes coded under each first-order code (right screen). If you have a large number of codes (i.e., >100), such as those included in this demonstration, you can select any or all of the codes from the HTML file and copy-and-paste them to an Excel (or other application) file to perform further analysis or data aggregation (e.g., abstracting all second-level themes to a higher dimension, or so-called *aggregate dimension*).

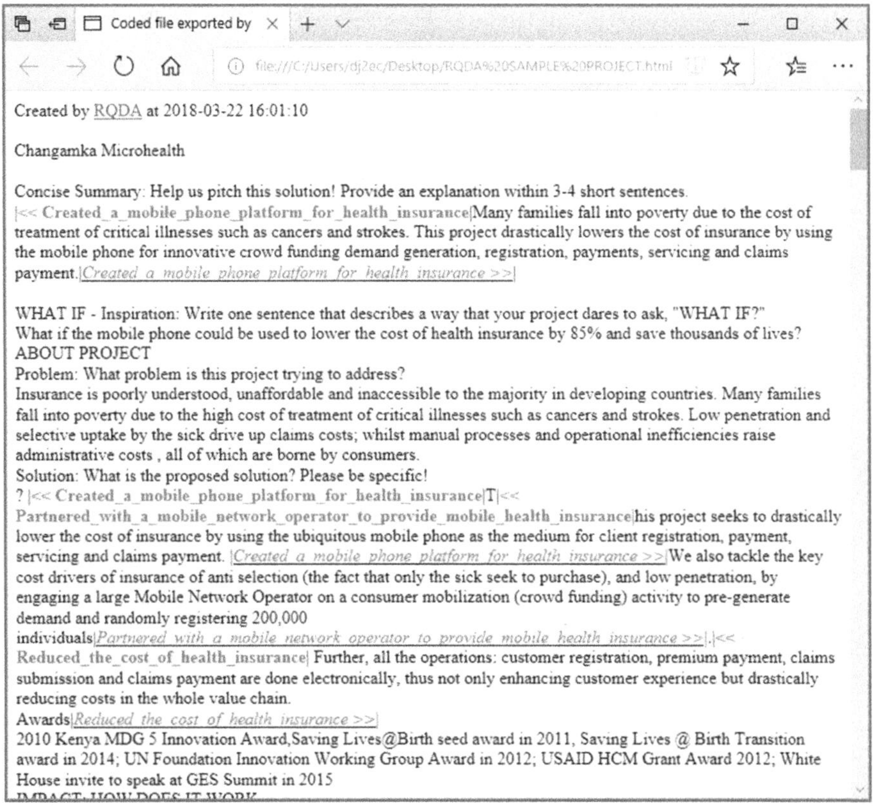

Fig. 10.12 Exported coded file in RQDA

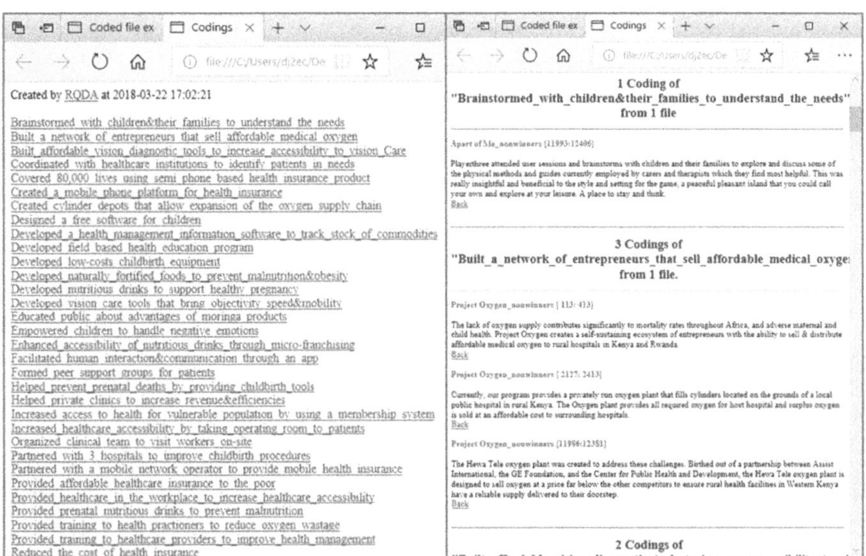

Fig. 10.13 Exported odings in HTML format

References

Gioia, D. A., Corley, K. G., & Hamilton, A. L. (2013). Seeking qualitative rigor in inductive research: Notes on the Gioia methodology. *Organizational Research Methods, 16*(1), 15–31.

Huang, R. (2014). RQDA: R-based qualitative data analysis. *R package version 0.2–7*. Retrieved on 18 April 2018, from, http://rqda.r-forge.r-project.org

Strauss, A., & Corbin, J. (1998). *Basics of qualitative research: Procedures and techniques for developing grounded theory*. Thousand Oaks: Sage.

Tracey, P., Phillips, N., & Jarvis, O. (2011). Bridging institutional entrepreneurship and the creation of new organizational forms: A multilevel model. *Organization Science, 22*(1), 60–80.

Chapter 11
Grounded Theory Development

Abstract This chapter discusses the technique used to transform the coded data into a "data structure"—by aggregating the first-order codes, second-order themes and finally the theoretical dimensions—and how to reconstruct them into a grounded process model or framework as the final outputs of qualitative research.

Keywords Grounded theory · Data structure · Theoretical dimensions · First-level code · Second-level theme · Grounded process model

11.1 Building Aggregate Theoretical Dimensions

As discussed above, as of April 2018, RQDA allows users to create only two code hierarchies, from first-order codes to second-order themes, so as to keep the data aggregation process simple. However, as our illustrative project demonstrates, the plotted codes map can look quite confusing when too many codes (>100) are created and plotted in RQDA. (As a side note, users can export RQDA-generated files to software such as SPSS or the R program for further statistical analysis). To allow aggregation up to the third-order codes (or creating *aggregate theoretical dimensions*), we copied all the first-order codes and second-order themes and pasted them into a spreadsheet (e.g., Excel) file, and then reorganized the sequence of codes based on their similarities and differences (Gioia et al. 2013) as shown in Table 11.1.

As shown in Table 11.1, we used similar colors to highlight first-order codes that fit with a particular second-order theme and used various colors to represent other first-order codes abstracted into other second-order themes. Finally, based on all second-order themes created, we developed the third-order or aggregate *theoretical dimensions* categories. In this aggregation process, we moved back and forth between the first-order codes, second-order themes, aggregate dimensions, and the literature/theory to find, generate or position new concepts from the data. This approach enables researchers to create multi-order structures of categories, thus making it easier for readers and gatekeepers to understand and trace both the infor-

Y. Chandra, L. Shang, *Qualitative Research Using R: A Systematic Approach*,
https://doi.org/10.1007/978-981-13-3170-1_11

Table 11.1 Gioia's data structure categorized in colours using excel

First Level Codes	Second Level Codes	Thrid Level Categories
Designed_a_free_software_for_children	Provided affordable healthcare services	Improving healthcare affordability
Built_a_network_of_entrepreneurs_that_sell_affordable_medical_oxygen	Provided affordable healthcare products	
Built_affordable_vision_diagnostic_tools_to_increase_accessibility_to_vision_Care		
Developed_low-costs_childbirth_equipment		
Created_a_mobile_phone_platform_for_health_insurance	Developed affordable health insurance	
Covered_80,000_lives_using_semi_phone_based_health_insurance_product		
Provided_affordable_healthcare_insurance_to_the_poor		
Reduced_the_cost_of_health_insurance		
Used_lung-sum_cash_payout_for_insurance_claims		
Increased_access_to_health_for_vulnerable_population_by_using_a_membership_system	Developed a healthcare membership system	
Designed_a_free_software_for_children	Use of Information Communication Technology	Improving healthcare accessibility
Partnered_with_a_mobile_network_operator_to_provide_mobile_health_insurance		
Used_interactive_games_to_enhance_children's_emotional_well being		
Facilitated_human_interaction&communication_through_an_app		
Provided_healthcare_in_the_workplace_to_increase_healthcare_accessibility	Approached patients actively	
Organized_clinical_team_to_visit_workers_on-site		
Increased_healthcare_accessibility_by_taking_operating_room_to_patients		
Brainstormed_with_children&their_families_to_understand_the_needs		
Used_mobile_surgery_to_deliver_surgical_care_in_remote_locations	Delivered healthcare services in romote areas	
Organized_clinical_team_to_visit_workers_on-site		
Provided_healthcare_in_the_workplace_to_increase_healthcare_accessibility		
Provided_training_to_health_practioners_to_reduce_oxygen_wastage	Provided training to healthcare providers	Improving healthcare procedure
Provided_training_to_healthcare_providers_to_improve_health_management		
Used_mobile_surgery_to_deliver_surgical_care_in_remote_locations	Delivered healthcare services in remote areas	
Organized_clinical_team_to_visit_workers_on-site		
Provided_healthcare_in_the_workplace_to_increase_healthcare_accessibility		
Developed_field_based_health_education_program	Provided public healthcare education	
Educated_public_about_advantages_of_moringa_products		
Formed_peer_support_groups_for_patients	Provided peer support	Empowering patients
Facilitated_human_interaction&communication_through_an_app	Empowered patients to enhance emotional well-being	
Used_interactive_games_to_enhance_children's_emotional_well being		
Empowered_children_to_handle_negative_emotions		
Coordinated_with_healthcare_institutions_to_identify_patients_in_needs	Partnered with healthcare institutions	Partnering with Cross-sector Organizations
Helped_private_clinics_to_increase_revenue&efficiencies		
Partnered_with_3_hospitals_to_improve_childbirth_procedures		
Provided_training_to_health_practioners_to_reduce_oxygen_wastage		
Provided_training_to_healthcare_providers_to_improve_health_management		
Partnered_with_a_mobile_network_operator_to_provide_mobile_health_insurance	Partnered with business companies	

mants' verbal evidence and how the qualitative researchers linked them to the concepts and theories posited. For example, we observed that the SI projects' partnership strategies varied but could be aggregated into two second-order themes: *SIs partnered with health institutions* and *SIs partnered with business companies*. Therefore, we combined the variety of partnership strategies into ***Partnering with Cross-sectoral Organizations*** (see Table 11.1).

As shown in our project demonstration (Table 11.1), we aggregated 38 first-order codes into 14 second-order themes and finally aggregated the second-order themes into the following five aggregate theoretical dimensions: ***(1) improving healthcare affordability, (2) improving healthcare accessibility, (3) improving healthcare procedure, (4) empowering patients,*** and ***(5) partnering with cross-sectoral organizations.***

Alternatively, users can also use software such as Microsoft Visio to draw the data structure (see Fig. 11.1 below).

In efforts to understand how SI projects promote social changes, we developed a grounded process model that depicts the interrelationships between the strategies used by SI projects and the outcomes (see Fig. 11.2 below). Based on the codings, we found that social innovators' strategies to address social problems can be explained using three perspectives: **market-oriented**, **human-oriented** and **institution-oriented**. By adopting these strategies, the social innovators have created immediate social outcomes such as **higher medical insurance coverage**, **more access to healthcare services in rural areas,** and **increased public health awareness**, which eventually leads to **improved health indicators** among target beneficiaries.

Our codes aggregation above proceeded *iteratively* (not linearly) as suggested by Gioia et al. (2013). The process of demonstrating why and how (both existing and new) concepts emerged enhances the *transparency* and *trustworthiness* of qualitative research and facilitates communication among research collaborators, journal reviewers, and editors.

11.2 Advanced Analysis Using R

A powerful feature of RQDA is that it enables researchers to perform complex queries (e.g., Boolean searches) on coded data by typing syntax into the R console. RQDA can also be combined with *tm*, a text-mining package in R, to perform text mining. This function helps qualitative researchers to further manage their data and perform multi-method research on a single platform. Thus, it meets the needs of various users and project types. Users can perform some simple queries and additional functions in R based on the following syntax samples: http://www.inside-r.org/packages/cran/RQDA.

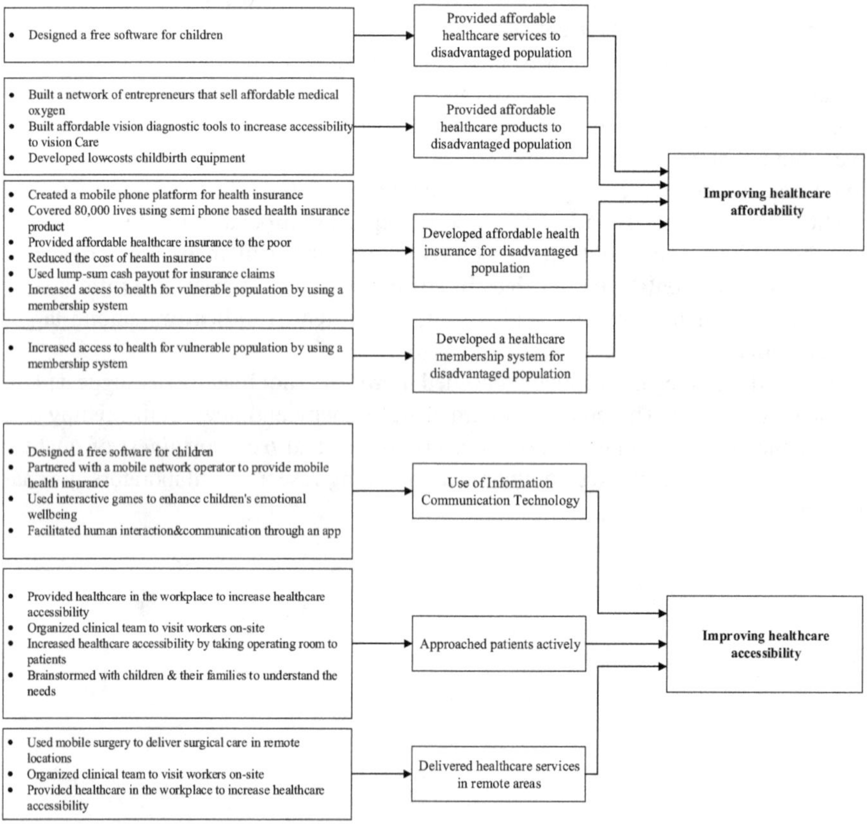

Fig. 11.1 A data structure for a sample project (drawn using Visio)

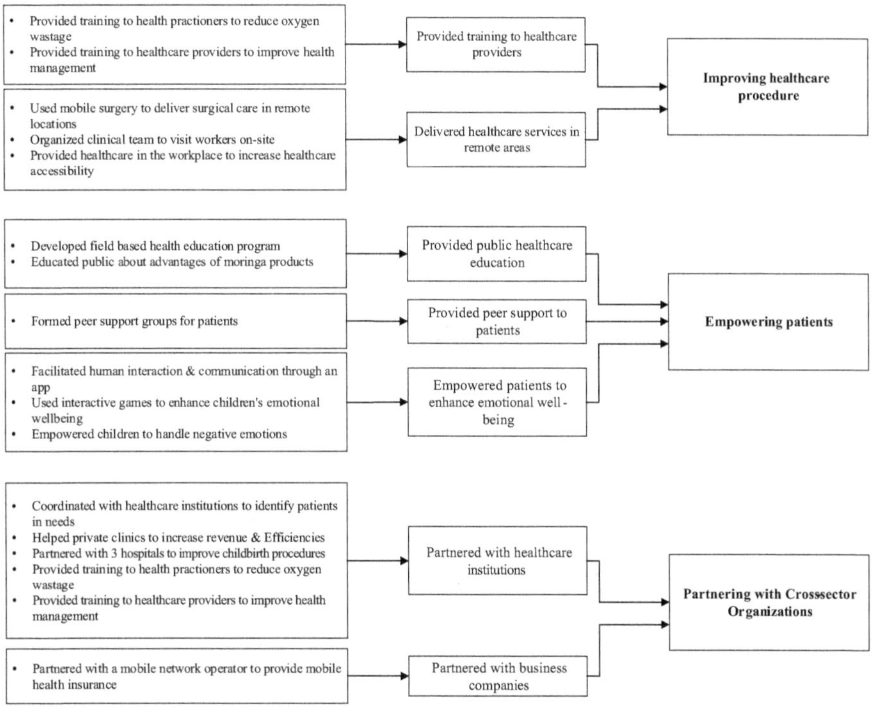

Fig. 11.1 (continued)

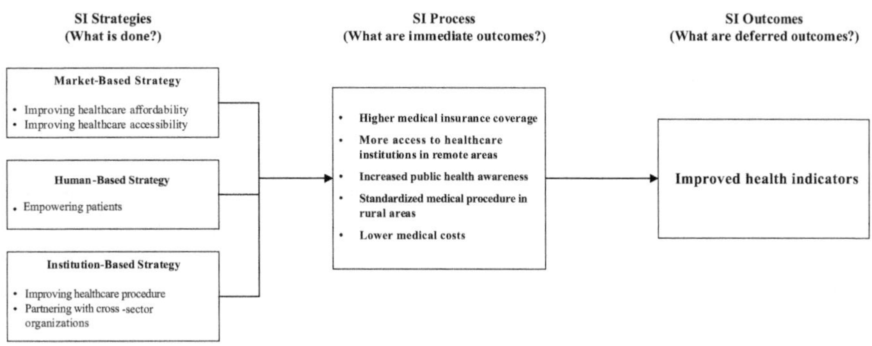

Fig. 11.2 A grounded process model of social innovation in the healthcare sector

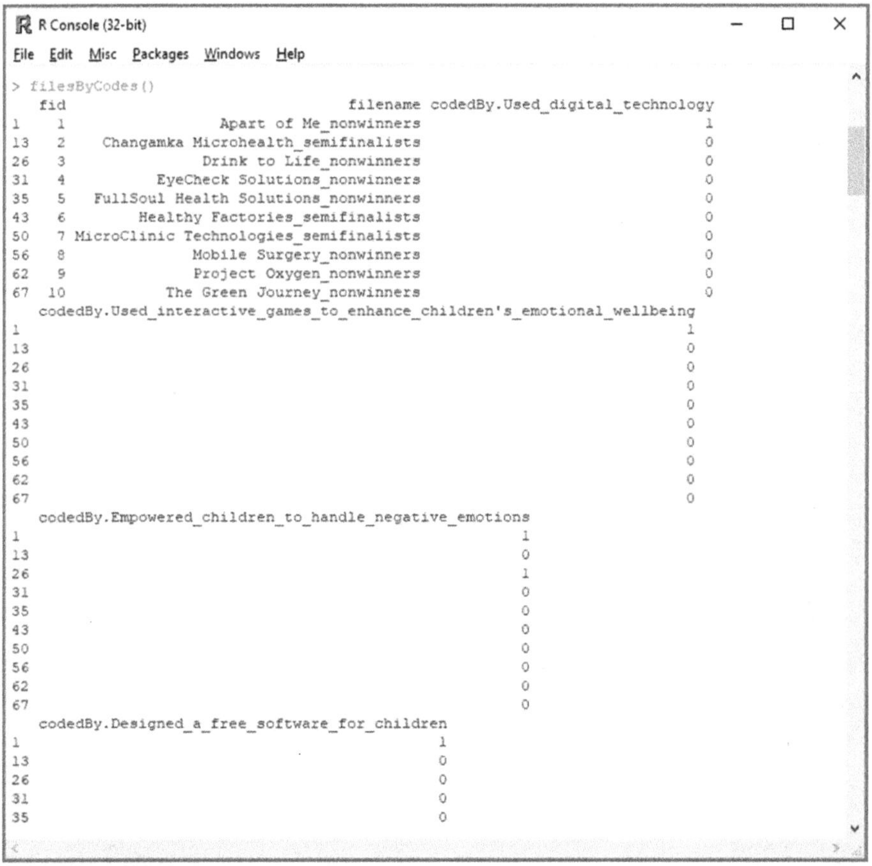

Fig. 11.3 A data frame of codes in RQDA

For example, if users want to see which codes are associated with each file, they can type *filesByCodes()* in the R console. Then, they can select codes from the pop-up window, and data frames appear as the search result (see Fig. 11.3).

Users can also create a summary of codings by typing *summaryCodings()*. By so doing, a list of the coding summary will detail information about the number of codings for each code, the average number of characters in the codings for each code and the number of files coded for each code (see Fig. 11.4).

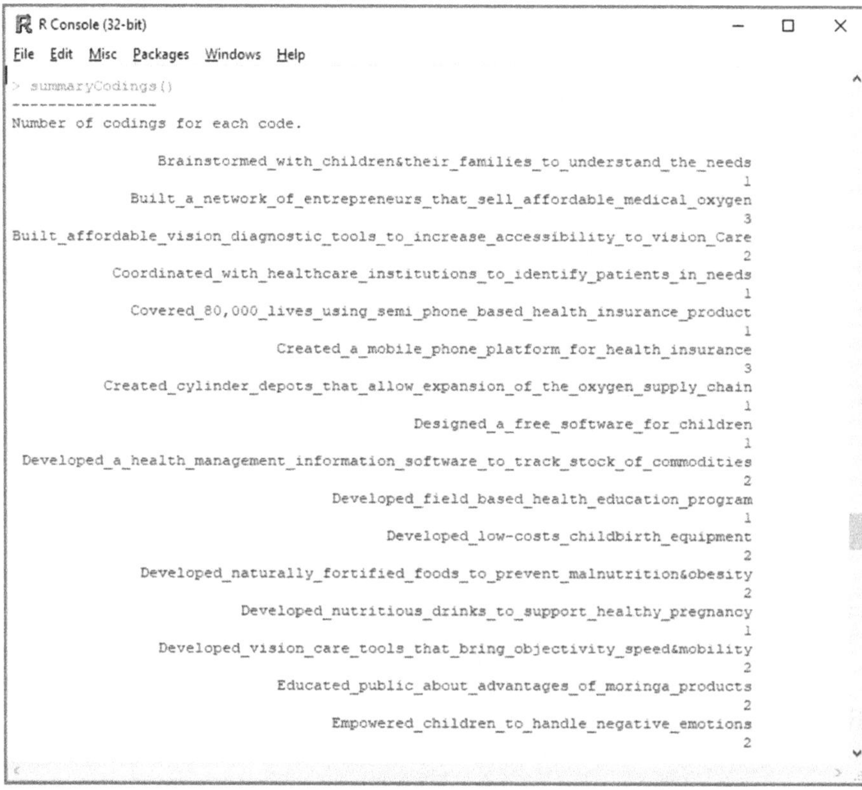

Fig. 11.4 A summary of codings in RQDA

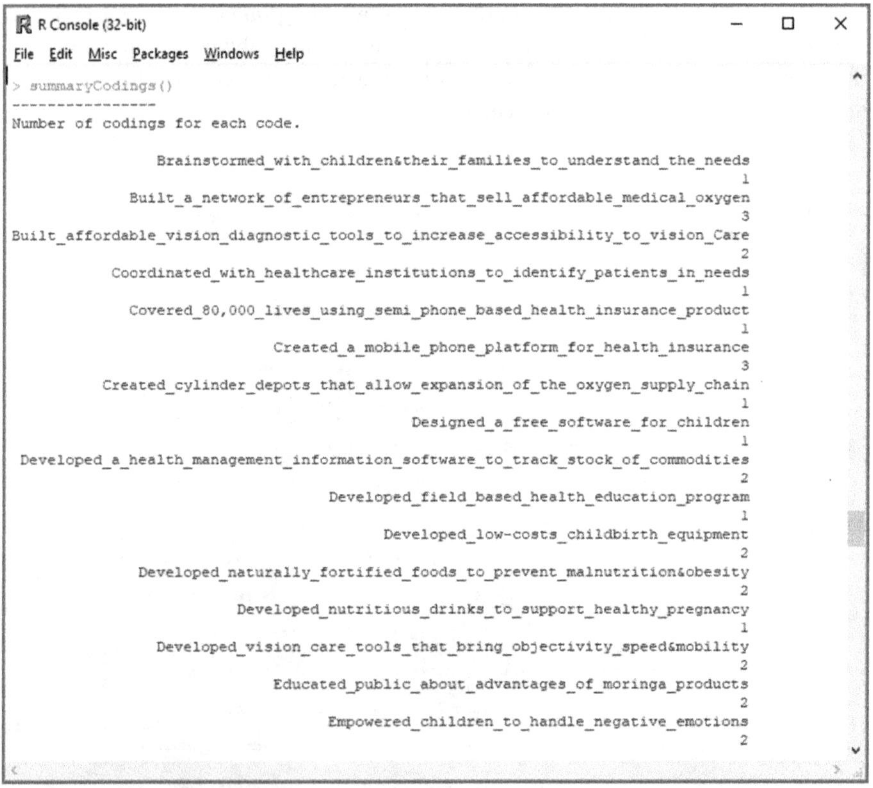

R Console (32-bit)

File Edit Misc Packages Windows Help

```
> summaryCodings()
----------------
Number of codings for each code.

                Brainstormed_with_children&their_families_to_understand_the_needs
                                                                                1
             Built_a_network_of_entrepreneurs_that_sell_affordable_medical_oxygen
                                                                                3
Built_affordable_vision_diagnostic_tools_to_increase_accessibility_to_vision_Care
                                                                                2
           Coordinated_with_healthcare_institutions_to_identify_patients_in_needs
                                                                                1
             Covered_80,000_lives_using_semi_phone_based_health_insurance_product
                                                                                1
                             Created_a_mobile_phone_platform_for_health_insurance
                                                                                3
          Created_cylinder_depots_that_allow_expansion_of_the_oxygen_supply_chain
                                                                                1
                                          Designed_a_free_software_for_children
                                                                                1
 Developed_a_health_management_information_software_to_track_stock_of_commodities
                                                                                2
                                 Developed_field_based_health_education_program
                                                                                1
                                       Developed_low-costs_childbirth_equipment
                                                                                2
              Developed_naturally_fortified_foods_to_prevent_malnutrition&obesity
                                                                                2
                      Developed_nutritious_drinks_to_support_healthy_pregnancy
                                                                                1
                Developed_vision_care_tools_that_bring_objectivity_speed&mobility
                                                                                2
                             Educated_public_about_advantages_of_moringa_products
                                                                                2
                                   Empowered_children_to_handle_negative_emotions
                                                                                2
```

Fig. 11.4 (continued)

References

Gioia, D. A., Corley, K. G., & Hamilton, A. L. (2013). Seeking qualitative rigor in inductive research: Notes on the Gioia methodology. *Organizational Research Methods, 16*(1), 15–31.

Chapter 12
Conclusion

Abstract In this book, we outlined and discussed various dominant paradigms in qualitative research—the Gioia, Yin-Eisenhardt, and Langley approaches—and demonstrated the abilities of computer-assisted qualitative research software (CAQDAS) particularly RQDA to extend and enrich the Gioia approach. We demonstrated how the fusion of R capabilities and the Gioia approach enhanced the *trustworthiness* and *rigor* of qualitative research and improved communication among researchers and gatekeepers (journal reviewers and editors). We believe that this systematic, transparent, trustworthy, creative yet disciplined approach to qualitative theorizing can help *legitimize* qualitative research and place it on par with other types of research including variance-based, simulation-based, experiment-based, and mathematics-based methodologies.

Achieving a consensus on how best to conduct qualitative research may prove to be illusive, and is beyond the scope of this book. But an appreciation of the strengths and limitations of each qualitative research approach will help researchers find an appropriate approach to suit individual needs, purposes and the type and amount of data necessary.

We demonstrated the application of the R-based Gioia approach—using a six-step procedure. The procedure should not be viewed as a rigid formula, but rather a guide to assist and even inspire researchers. We embraced the Gioia approach because it fits well with the nature of the data that we are working with (i.e., interview texts, organizational profiles, news articles), the kind of questions qualitative researchers ask (i.e., why and how, or process questions), and the types of contributions qualitative researchers hope to make. We believe that the approach explicated in this book can also be applied to the Yin-Eisenhardt approach (focused on confirming or refuting *variables*) and the Langley approach (focused on *events* and how they unfold) for qualitative research. The R platform and **RQDA are only tools** and their power lie in the philosophy and skills of the users.

One major advantage of RQDA is that it is a *free* tool, an extension of the powerful R computing platform that offers access to various data management, computing, and data presentation/visualization capabilities. As a free CAQDAS tool,

Y. Chandra, L. Shang, *Qualitative Research Using R: A Systematic Approach*,
https://doi.org/10.1007/978-981-13-3170-1_12

RQDA allows junior and senior researchers to take "baby steps" before investing in proprietary (and sometimes costly) suites of textual data analysis programs. RQDA is GUI based and is therefore easy to use for both non-programmers and novice researchers. First-time users can copy and paste the syntax provided in this book to experiment with and master RQDA; therefore, they need no programming experience. RQDA supports textual materials in English and more than a dozen non-English languages, thus facilitating cross-cultural research. Moreover, RQDA is supported by SQLite (a relational database management system), which enables additional data-management functionality by writing R syntax.

Nevertheless, RQDA has some shortcomings. One major weakness is that it supports only plain-text (.txt) files. It does not support audio files, videos, spreadsheets, tables, photos and other non-plain text formats that are available in the popular, proprietary NVivo and ATLAS.ti programs. However, most qualitative research projects, focus on analyzing textual materials (particularly interview data, organizational documents, and news articles); therefore, this weakness is not a major deficiency for qualitative researchers. Another issue is that as an open-source platform, the RQDA developer does not offer users any technical support. Thus, users who prefer dedicated support for their CAQDAS platform may prefer NVivo and ATLAS.ti. Nevertheless, there are ample opportunities to increase the interaction among both experienced and novice RQDA users via online chat forums, presentations and "coaching programs" at major conferences, in addition to direct correspondence with experienced users (e.g. RQDA video tutorials: https://www.youtube.com/watch?v=mLsyGH3ztYY; R developers/users forums, Github: https://github.com/Ronggui/RQDA, Stack Overflow: https://stackoverflow.com/questions/43726838/rqda-cannot-load-rgtk2-so; R-related blog, R-bloggers: https://www.r-bloggers.com/qualitative-text-analysis-in-r-with-rqda/). This book fills in the 'support gap' above and will hopefully be a useful guide for readers.

Finally, RQDA enables users to combine qualitative and quantitative data analysis through freely available R packages (you may find a list of R packages from https://cran.r-project.org/web/packages/available_packages_by_name.html), thus enabling *mixed-methods research*. RQDA supports qualitative and quantitative *content analysis*–style research on textual data (e.g., policy documents, news articles, and other online, publicly available textual data such as Facebook and Twitter posts, blogs; Chandra et al. 2016; Perry and Bodkin 2000). RQDA can also be combined with "bag of words" software, such as Linguistic Inquiry and Word Count (LIWC) or DICTION, which converts textual data into numbers for statistical analysis (Chandra 2016; Short and Palmer 2008). The R-based Gioia approach explicated in this book can be fruitfully combined with quantitative research such as *surveys* or (lab or field) ***experiments*** in which insights from quantitative data can be corroborated and pattern-matched with in-depth, concrete verbal data from informants to enable both data and method triangulation.

We hope that this book will serve as a catalyst to further refine, improve and legitimize qualitative scholarship, particularly through the use of the R package, RQDA.

References

Chandra, Y. (2016). A rhetoric-orientation view of social entrepreneurship. *Social Enterprise Journal, 12*(2), 161–200.

Chandra, Y., Jiang, L. C., & Wang, C. J. (2016). Mining social entrepreneurship strategies using topic modeling. *PLoS One, 11*(3), e0151342.

Perry, M., & Bodkin, C. (2000). Content analysis of Fortune 100 company Web sites. *Corporate Communications: An International Journal, 5*(2), 87–97.

Short, J. C., & Palmer, T. B. (2008). The application of DICTION to content analysis research in strategic management. *Organizational Research Methods, 11*(4), 727–752.

Index

Y. Chandra, L. Shang, *Qualitative Research Using R: A Systematic Approach*,
https://doi.org/10.1007/978-981-13-3170-1